Wave Asymptotics

Participants of the meeting.

Fritz Ursell lecturing.

Wave Asymptotics

The proceedings of the meeting to mark the retirement of Professor Fritz Ursell from the Beyer Chair of Applied Mathematics in the University of Manchester

Edited by

P. A. Martin
G. R. Wickham

Department of Mathematics
University of Manchester

CAMBRIDGE
UNIVERSITY PRESS

Published by the Press Syndicate of the University of Cambridge
The Pitt Building, Trumpington Street, Cambridge CB2 1RP
40 West 20th Street, New York, NY 10011-4211, USA
10 Stamford Road, Oakleigh, Victoria 3166, Australia

First Published 1992

Printed in Canada

Library of Congress Cataloging-in-Publication Data

Wave asymptotics : the proceedings of the meeting to mark the
retirement of professor Fritz Ursell from the Beyer Chair of Applied
Mathematics in the University of Manchester / edited by P. A. Martin,
G. R. Wickham.

p.   cm.

Held March 1990.
Includes bibliographical references.

ISBN 0-521-41414-8 (hard cover)

1. Water waves – Congresses.   2. Differential equations. Linear –
Asymptotic theory – Congresses.   I. Ursell, Fritz, 1923–   .
II. Martin, P. A.   III. Wickham, G. R.
TC172.W38   1992
27'.042 – dc20   91–36847
                                                            CIP

A catalog record for this book is available from the British Library.

ISBN 0-521-41414-8 hardback

Contents

viii   Contents

Foreword

Professor Fritz Ursell's earliest association with the University of Manchester began in 1947 when he was awarded an ICI Research Fellowship in Applied Mathematics, but his principal period of service dates from 1961 when he joined the Department of Mathematics as the Beyer Professor. His path towards this point in his career was not always straightforward. He was born in Düsseldorf, Germany in 1923; his father was a doctor. At first he was educated in German schools, but in 1936, with the situation of German Jews becoming steadily more intolerable and with no possibility of a university education, he was sent to Clifton College (near Bristol). He remained at Clifton until his progress was further disrupted by the outbreak of war; in 1940 the British Government ruled that German refugees could not live within 50 miles of the coast, and so he completed his sixth-form studies at Marlborough College in Wiltshire, from where he obtained a Major Scholarship to Trinity College Cambridge. He graduated from Cambridge with a First in Mathematics Part II in 1942 and a Distinction in Part III in 1943.

Fritz's first employment was with the British Admiralty during the war, and soon he was made part of Group W (the Wave group), whose task it was to determine the rules for forecasting ocean waves and surf for landings due to take place in the Pacific. The fact that the Admiralty were prepared to employ on secret work someone still classified as an enemy alien is a tribute to their good judgement; it also shaped the course of Fritz's subsequent career as a mathematical researcher in the linear theory of water waves. In 1947 he was made an ICI Research Fellow in Applied Mathematics at Manchester, joining Sydney Goldstein and James Lighthill. He left in 1950 to take up appointments in Cambridge, but returned to Manchester in 1961 to take the Beyer Chair of Applied Mathematics, succeeding James Lighthill.

For almost three decades Fritz Ursell has maintained his, and Manchester's, position in the forefront of mathematical research in the linear theory of water waves and the theory of asymptotics. Many of his publications are landmarks in the literature, and he has frequently attracted distinguished overseas researchers to Manchester. He has been a prolific supervisor of research students, many of whom now occupy senior positions in academic institutions throughout the world. His academic achievements were officially recognised in 1972 when he was elected a fellow of the Royal Society.

Within the Department, Professor Ursell has had a leading role. Generations of undergraduates and postgraduates alike have been impressed by his mathematical skills and his patience in explaining the more delphic points. While not seeking to be 'in administrative charge' he was ready to do his stint and is remembered for having a greater amount of administrative skill than he gave himself credit for. As a member of Senate, and of Court, he is well known for speaking out, always with calmness and reason, on issues which he regards as matters of principle; this has been particularly evident when he perceived that freedoms or due process were being undermined.

Although Fritz Ursell has now officially retired, he has not given up his research. He will continue to be an important contributor in the Mathematics Department, assisting his colleagues with his renowned knowledge of mathematical methods and publishing his own work; long may he remain one of the Department's finest assets.

Professor R. Douglas Gregory
Department of Mathematics
University of Manchester
Manchester M13 9PL
U.K.

Preface

At the end of September 1990, Fritz Ursell retired from the Beyer Chair of Applied Mathematics, having occupied it since 1961. This Chair, in the Department of Mathematics in the University of Manchester, had previously been held by Horace Lamb, Sydney Goldstein and James Lighthill.

To mark his retirement, a two-day meeting was held in Manchester in March, 1990. There were invited lectures from eleven scientists, all of whom have interacted fruitfully with Fritz over the years. The idea was to have lectures covering each of his areas of influence by people who know him well. The twelfth lecture was given by Fritz himself, in which he described some problems that he had *not* solved! There was also a dinner in his honour, at which the speakers were Frank Olver and Richard Holford. The meeting was attended by over a hundred people, including many of his former Ph.D. students.

Fritz Ursell has made many seminal contributions to applied mathematics in general. In particular, his contributions to the linear theory of water waves and to the development of rigorous asymptotic methods are especially valuable and well known. These observations led Sir James Lighthill to suggest the title for this volume, namely *Wave Asymptotics*.

This volume contains papers from all the lecturers at the meeting. It also includes a list of Fritz Ursell's publications, together with his own commentary. Of course, this list is still growing; although Fritz has now retired, he continues to research, unhindered by any administrative burdens. We wish him well for the future.

It is our pleasure to thank all the contributors for their chapters and for making the meeting such a success. We also thank Douglas Gregory, the current Head of the Department of Mathematics and one of Fritz's former Ph.D. students, for his Introduction. The volume itself was compiled in Manchester using LaTeX from a wide variety of manuscripts. Most were typed or retyped by Gail Buckley; we are very grateful for her excellent assistance. We also acknowledge the encouragement provided by Alan Harvey of Cambridge University Press.

P.A. Martin
G.R. Wickham
March 1991

List of contributors

J.P. Breslin, Ciudad de Las Communicaciones, 11–3,
03193 San Miguel de Salinas, Alicante, Spain.

J.N.L. Connor, Department of Chemistry, University of Manchester,
Manchester M13 9PL, England.

K. Eggers, Institut für Schiffbau, Lämmersieth 90, 2000 Hamburg,
Germany.

D.V. Evans, School of Mathematics, University of Bristol,
Bristol BS8 1TW, England.

V. Hutson, Department of Applied and Computational Mathematics,
University of Sheffield, Sheffield S10 2TN, England.

F.G. Leppington, Department of Mathematics, Imperial College,
London SW7 2BZ, England.

J. Lighthill, Department of Mathematics, University College London, London WC1E 6BT, England.

R.E. Meyer, Center for the Mathematical Sciences, University of Wisconsin, Madison, Wisconsin 53705, U.S.A.

J.N. Newman, Department of Ocean Engineering, Massachusetts Institute of Technology, Cambridge, Massachusetts 02139, U.S.A.

F.W.J. Olver, Institute for Physical Science and Technology, University of Maryland, College Park, Maryland 20742, U.S.A.

E.O. Tuck, Department of Applied Mathematics, University of Adelaide, GPO Box 498, Adelaide, SA 5001, Australia.

1

Asymptotic Behaviour of Anisotropic Wave Systems Stimulated by Oscillating Sources

Sir James Lighthill
University College London

1.1 Abstract

The theory of the far-field asymptotic behaviour of anisotropic waves, forced by a local distribution of sources of fixed frequency ω_0, expresses those waves at a given distant location in the form of an integral over just a part of the wavenumber surface S (the surface in wavenumber space that is specified, for frequency ω_0, by the dispersion relationship). In the theory as set out by Lighthill (1960, 1978) the integral is taken over S_+, defined as that part of S for which the dispersion relationship makes the gradient of frequency in wave-number space have a positive component (this is a form of "the radiation condition") in the direction of the given distant location.

On the other hand a recent more rigorous treatment (Lighthill 1990) has shown that the integral is more correctly taken over a combination of the open surface S_+ (defined as the above part of the real wavenumber surface S) and S_{-i} (defined as a part of the complex wavenumber surface obtained by proceeding from the edge C of S_+ in a negative pure-imaginary direction). The consequent suppression of any discontinuity at C is important, as eliminating all possibility of an associated spurious $O(x^{-1})$ asymptotic signal at large distances x from the source region.

The new rigorous analysis confirms the correctness of the results obtained (Lighthill 1978) by ignoring any possible contribution from C; in other words, by simply applying stationary-phase principles to the integral over S_+. At the end of the paper, some of these results are recalled briefly, together with their application to a particular problem of wave asymptotics in which a powerful interest has been taken by Ursell (1960) and also by Inui (1962): the analysis of ship waves and of approaches towards reducing the wavemaking resistance of ships.

1.2 Introduction

At this Retirement Meeting honouring a great exponent of fluid dynamics, of waves and of asymptotics, who since 1961 held Manchester University's Beyer Chair of Applied Mathematics, I happily seize an early opportunity to make a remark both on my own behalf and simultaneously, perhaps, for all those other former holders of the Beyer Chair (from Horace Lamb to Sydney Goldstein) who are with us no longer. In this brief preliminary remark I want to voice a view which (in some mystic sense!) I confidently claim as our collective view of how magnificently the traditions of this famous Chair have during the past three decades been maintained – in fluid dynamics, in waves, in asymptotics – by the eminent tenure of Fritz Ursell.

And next, like most of the speakers at this meeting, I proceed to offer a paper on a favourite theme (see title above) within the broad field that may be described as *Wave Asymptotics:* a paper in which, while referring to earlier treatments of my theme by Lighthill (1960, 1978), I comprehensively reconsider this theme's fundamentals. Indeed, in a certain sense I pay homage rather specifically to Fritz Ursell by reconsidering my theme with a highly necessary injection of *added rigour* !

I shall present, then, a reconsideration of the asymptotics of forced vibrations in anisotropic wave systems that are *homogeneous*; being described, indeed, by linear partial differential equations with constant coefficients. The systems are subjected to local forcing (specified by a right-hand side of the partial differential equation that is nonzero over just a limited region of space) at fixed radian frequency ω_0.

In relation to my choice of topic I may perhaps make four general points, as follows.

(a) Anisotropic wave propagation is widespread and important.

(b) It is easy (see §1.11 below) to extend the theory to the case of travelling forcing effects; where "Doppler shifts" make a problem effectively anisotropic even in cases where the wave system as such is isotropic.

(c) The theory is readily adapted (see §1.3 below) to cases of boundary forcing.

(d) The theory for homogeneous systems shows wave energy travelling away from the local forcing region along straight rays, but it can be extended to systems with gradually varying properties (leading to gradually varying coefficients of the partial differential equations) by the refraction theory for rays (see Lighthill 1978, section 4.5); which states that, in general linear systems with a dispersion relationship

$$\omega = \omega(k_i, x_i) \tag{1.1}$$

that links frequency ω to vector wavenumber k_i at the point with position vector x_i, the rays are described by equations in Hamiltonian form,

$$\frac{dx_i}{dt} = \frac{\partial \omega}{\partial k_i}, \frac{dk_i}{dt} = -\frac{\partial \omega}{\partial x_i}, \tag{1.2}$$

implying that wave energy travels at the group velocity $\partial \omega / \partial k_i$ appropriate to its wavenumber k_i, which itself varies gradually by refraction.

In the present lecture, however, I limit myself to homogeneous systems for which the dispersion relationship takes a simpler form,

$$\omega = \omega(k_i), \tag{1.3}$$

that allows wave energy to travel along straight rays with unchanging wavenumber.

1.3 Explanation of the Fourier integral which has to be asymptotically estimated

I consider local forcing by a source distribution $e^{i\omega_0 t} f(\mathbf{x})$, where the forcing amplitude $f(\mathbf{x})$ has Fourier transform $F(\mathbf{k})$ so that it can be written as a Fourier integral

$$f(\mathbf{x}) = \int F(\mathbf{k}) e^{-i\mathbf{k}\cdot\mathbf{x}} d\mathbf{k}, \tag{1.4}$$

taken over the space of all wavenumbers

$$\mathbf{k} = (k_1, k_2, k_3). \tag{1.5}$$

Then the mathematical problem which has to be solved is to estimate another Fourier integral,

$$q = e^{i\omega_0 t} \int \frac{F(\mathbf{k})}{B(\omega_0, \mathbf{k})} e^{-i\mathbf{k}\cdot\mathbf{x}} d\mathbf{k}, \tag{1.6}$$

in which the denominator B represents a function of ω and $\mathbf{k}$ such that the equation

$$B(\omega, \mathbf{k}) = 0 \tag{1.7}$$

is a form of the dispersion relationship (1.3).

The integral expression (1.6) is derived in three dimensions (the case here treated) from a partial differential equation

$$B\left(-i\frac{\partial}{\partial t}, i\frac{\partial}{\partial x_1}, i\frac{\partial}{\partial x_2}, i\frac{\partial}{\partial x_3}\right) q = e^{i\omega_0 t} f(\mathbf{x}) \tag{1.8}$$

$$= e^{i\omega_0 t} \int F(\mathbf{k}) e^{-i\mathbf{k}\cdot\mathbf{x}} d\mathbf{k},$$

where B is a polynomial – or, more generally, a rational function – in its four variables. Essentially this is because the Fourier transform $e^{i\omega_0 t} Q(\mathbf{k})$ of q must satisfy

$$B(\omega_0, \mathbf{k}) Q = F, \tag{1.9}$$

from which equation (1.6) follows immediately.

In two dimensions an identical integral expression can, needless to say, be obtained from an equation similar to (1.9) but with only the three independent variables t, x_1 and x_2. On the other hand, there is also a second, perhaps more interesting, way in which equation (1.6) may arise in two dimensions.

This is the case of boundary forcing at a plane boundary $x_3 = $ constant, where a certain quantity η is subjected to local forcing

$$\eta = e^{i\omega_0 t} f(x_1, x_2) = e^{i\omega_0 t} \int F(\mathbf{k}) e^{-i\mathbf{k}\cdot\mathbf{x}} d\mathbf{k} \qquad (1.10)$$

at frequency ω_0, the integration being over the space of all wavenumbers $\mathbf{k} = (k_1, k_2)$. Here, the three-dimensional system is supposed to satisfy a linear partial differential equation with constant coefficients and zero right-hand side; furthermore, the condition $\eta = 0$ (corresponding to the case when the boundary is free) yields the dispersion relationship (1.7). Finally, the quantity η (representing local forcing) is related to a basic independent variable $q(t, x_1, x_2)$ in such a way that

$$B(\omega_0, k_1, k_2)Q = F \qquad (1.11)$$

for the Fourier transform $e^{i\omega_0 t}Q(\mathbf{k})$ of q; a fact which directly yields the integral expression (1.6) for q.

At a meeting in Fritz Ursell's honour it is apposite to use *water waves* (say, gravity waves on the surface of deep water) to illustrate boundary forcing. If q is surface elevation then η, the surface pressure (relative to atmospheric), is given on linear theory as

$$\eta = -\rho g q - \rho \partial\phi/\partial t \qquad (1.12)$$

where ρ is fluid density and where the velocity potential ϕ satisfies a boundary condition

$$\partial q/\partial t = \partial\phi/\partial x_3 \qquad (1.13)$$

at the free surface's undisturbed position. On the other hand, Laplace's equation for motions in deep water (which vanish as $x_3 \to -\infty$) implies that waves of horizontal wave number (k_1, k_2) must everywhere satisfy

$$\partial\phi/\partial x_3 = -\left(k_1^2 + k_2^2\right)^{1/2}\phi; \qquad (1.14)$$

which, with equations (1.12) and (1.13), implies that the Fourier transform of q satisfies equation (1.11) with

$$B(\omega, k_1, k_2) = \rho\omega^2(k_1^2 + k_2^2)^{-1/2} - \rho g. \qquad (1.15)$$

Then, as expected, the equation (1.7) representing the case of a free boundary is a form of the classical dispersion relationship

$$\omega = (gk)^{1/2} = g^{1/2}(k_1^2 + k_2^2)^{1/4}; \qquad (1.16)$$

on the other hand, the forced boundary condition (1.10) describes the application of oscillatory pressure sources at the surface which generate outgoing waves given by the integral expression (1.6).

1.4   Special problems in the estimation of this Fourier integral

Whether in two or three dimensions there is an important and famous difficulty in estimating the Fourier integral (1.6). This is that the integral is mathematically ambiguous whenever the denominator $B(\omega_0, \mathbf{k})$ possesses zeros for real $\mathbf{k}$: zeros which, in general, are poles of the integrand[1].

Thus, whenever the integral with respect to k_1 (say) encounters such a pole for real k_1, we are constrained to ask whether we should take the integral to the right or to the left of the pole; or, alternatively, to use neither version exclusively but a linear combination of both. (The most famous such linear combination – which, I must however emphasize, does not prove to be the correct answer in the present problem – is the Cauchy Principal Value; namely, the arithmetic mean of the two versions.)

This mathematical ambiguity has, as is well known, a physical counterpart. A wave system, compatible with the given forcing which acts within a local region, might in principle be found to incorporate also free waves with their energy coming in "from infinity". (In a linear system, the linear combination of such free waves with the waves that are truly forced by the local sources would still satisfy the same partial differential equation.)

On physical grounds, however, we need to use that unique mathematical determination of the Fourier integral (1.6) which completely excludes any such free waves with energy coming in from infinity. Lighthill (1960) introduced a physically very natural approach to this determination of the waves truly and exclusively forced by the local sources when he proposed that the problem be replaced by an initial-value problem; namely, that of finding the waves in an initially undisturbed system excited by a local forcing that has gradually grown up from zero to its present value.

Specifically, he proposed a substitution of the local forcing function

$$e^{i\omega_0 t} f(\mathbf{x}) \quad \text{by} \quad e^{\epsilon t} e^{i\omega_0 t} f(\mathbf{x}) = e^{i(\omega_0 - i\epsilon)t} f(\mathbf{x}) \tag{1.17}$$

where ϵ is a small positive number. The substitution (1.17) allows the sources to have grown exponentially (from zero when $t = -\infty$) at a slow rate ϵ which may subsequently be allowed to tend to zero.

It might of course have been thought more natural to give the sources a "sudden switch-on" at time $t = 0$ and determine the subsequent behaviour for large t; the substitution (1.17) for the source function would then become a change to $H(t)e^{i\omega_0 t} f(\mathbf{x})$ where $H(t)$ is the Heaviside unit function. However, because the spectrum of $H(t)e^{i\omega_0 t}$ involves essentially all frequencies – including frequencies at which the system may possibly respond to forcing unstably, or in some other way different from its

---

[1] Essentially, this happens whenever the dispersion relationship (1.7) has solutions for $\omega = \omega_0$ representing unattenuated waves

response when sources at frequency ω_0 excite it – the approach to our estimation problem via a "sudden switch-on" involves grave inconveniences; some of which are familiar already in the relatively simple case of gravity waves on deep water from the intricate complexities of the Cauchy-Poisson problem.

We should also reject any approach to the estimation of the Fourier integral (1.6) which simply assumes that the ray theory (1.2) must asymptotically be satisfied; as is taken for granted in certain ways of applying "a radiation condition". Our aim, rather, should be to demonstrate the existence of an asymptotic behaviour consistent with ray theory from a rigorous estimate of the integral.

After rejecting those two alternative procedures, then, we return to the substitution (1.17) which, essentially, replaces ω_0 by $\omega_0 - i\epsilon$ so that the integral expression (1.6) which has to be asymptotically estimated becomes

$$q = e^{i(\omega_0 - i\epsilon)t} \int \frac{F(\mathbf{k})}{B(\omega_0 - i\epsilon, \mathbf{k})} e^{-i\mathbf{k}\cdot\mathbf{x}} d\mathbf{k}. \tag{1.18}$$

The modified integral expression (1.18) is in general unambiguously defined for small positive ϵ while its limiting value as ϵ becomes vanishingly small represents the waves actually generated in the local forcing region.

1.5   The estimation method of Lighthill (1960, 1978) briefly recapitulated

The method introduced by Lighthill (1960) for estimating the integral (1.18) utilises an early switch to new Cartesian axes, chosen so as to facilitate the calculation. Then, when estimation is completed, the results are thrown into a form invariant under such change of axes, so that they are valid equally in the original system of axes.

Specifically, with an origin in the midst of the source region, the method studies an arbitrary line ℓ stretching radially out from that origin and seeks to estimate the integral on ℓ at large distances from the origin. To this end, new axes are chosen with ℓ as the positive x_1-axis. At a point on ℓ, then, we have $x_2 = x_3 = 0$ so that $\mathbf{k}\cdot\mathbf{x} = k_1 x_1$; accordingly, the integral expression (1.18) becomes

$$q = e^{i(\omega_0 - i\epsilon)t} \int_{-\infty}^{\infty}\int_{-\infty}^{\infty} dk_2 dk_3 \int_{-\infty}^{\infty} \frac{F(\mathbf{k})e^{-ik_1 x_1} dk_1}{B(\omega_0 - i\epsilon, \mathbf{k})} \tag{1.19}$$

which must be estimated as $x_1 \to +\infty$. (In a two-dimensional problem, the k_3 integration in (1.19) would be suppressed.)

In equation (1.19) the inner integral

$$\int_{-\infty}^{\infty} \frac{F(\mathbf{k})e^{-ik_1 x_1} dk_1}{B(\omega_0 - i\epsilon, \mathbf{k})} \tag{1.20}$$

is estimated by lowering the path of integration in the complex k_1-plane (Figure 1.1) so as to give the imaginary part of k_1 a negative value

$$\text{Im } k_1 = -\kappa_1. \tag{1.21}$$

Even if κ_1 be of modest magnitude, the integral (1.20) along the new path (1.21) includes the factor $e^{-\kappa_1 x_1}$ which becomes very small as $x_1 \to +\infty$; when, indeed, the modulus of (1.20) must be less than or equal to the integrand of the integrand's modulus, namely

$$e^{-\kappa_1 x_1} \int_{-i\kappa_1-\infty}^{-i\kappa_1+\infty} \frac{|F(\mathbf{k})|d\kappa_1}{|B(\omega_0 - i\epsilon, \mathbf{k})|} = O(e^{-\kappa_1 x_1}). \tag{1.22}$$

Now Cauchy's theorem tells us that the difference between the integral (1.20) and the same integral taken along this different path (1.21) – so as to give it such an $O(e^{-\kappa_1 x_1})$ value – is equal to

$$(-2\pi i) \times \text{sum of residues at poles with } -\kappa_1 < \text{Im } k_1 < 0. \tag{1.23}$$

Thus, with an $O(e^{-\kappa_1 x_1})$ error, the inner integral (1.20) is given by the expression (1.23); where the first factor is $(-2\pi i)$ because it is in the negative sense that the difference integral encircles those poles.

In his calculation of expression (1.23), Lighthill (1960, 1978) confined his consideration to poles which for $\epsilon = 0$ are on the real k_1-axis but are

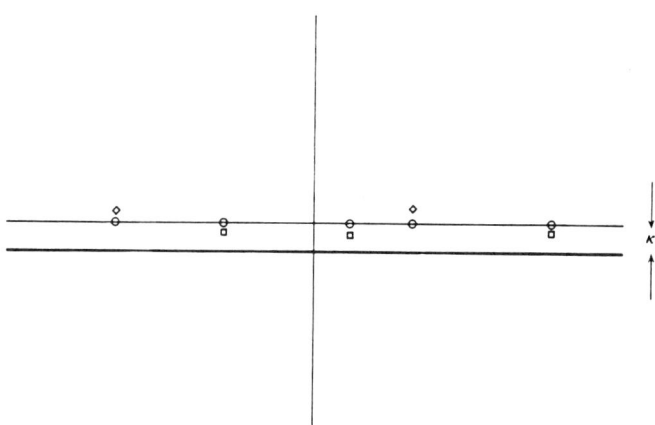

Figure 1.1: A lowered path of integration (thick line) in the complex k_1-plane for the inner integral (1.20). $\circ$ Positions of poles when $\epsilon = 0$; $\square$ Displaced position, when ω_0 is replaced by $\omega_0 - i\epsilon$ with $\epsilon > 0$, of poles k_1 such that $\partial\omega/\partial k_1 > 0$; $\diamond$ Displaced position of poles k_1 such that $\partial\omega/\partial k_1 < 0$. Only the poles $\square$ contribute a residue term when the path is lowered.

shifted off it for $\epsilon = 0$. Although a reconsideration of this issue will be given in §1.6, we here outline the analysis based on that original approach.

Lighthill (1960, 1978) pointed out that a solution of $B(\omega_0, \mathbf{k}) = 0$ for k_1 real becomes (see Figure 1.1) a solution of $B(\omega_0 - i\epsilon, \mathbf{k}) = 0$ with $\mathrm{Im}\, k_1 < 0$ if and only if the dispersion relationship $B(\omega, \mathbf{k}) = 0$ for $\mathbf{k}$ real defines a function $\omega = \omega(\mathbf{k})$ such that

$$\partial\omega/\partial k_1 > 0. \tag{1.24}$$

To this mathematically evident statement, the physical counterpart is that the only waves with frequency ω_0 that are expected to appear are those whose energy flow velocity has a positive component $\partial\omega/\partial k_1$ along the x_1-axis (implying that they could have been generated in the source region rather than coming in "from infinity").

In the limit as $\epsilon \to 0$, then, the expression (1.23) becomes $(-2\pi i)$ times the sum of residues of the inner integrand (1.20) at those poles which for $\epsilon = 0$ are on the real k_1-axis and where also the inequality (1.24) is satisfied. From any such pole, that is from a value of k_1 such that

$$\omega(\mathbf{k}) = \omega_0 \quad \text{and} \quad \partial\omega/\partial k_1 > 0, \tag{1.25}$$

the contribution to the inner integrand is

$$\frac{(-2\pi i) F(\mathbf{k})}{\partial B(\omega_0, \mathbf{k})/\partial k_1} e^{-ik_1 x_1} \tag{1.26}$$

by the usual rule for calculating a residue.

Finally, then, with an error $O(e^{-\kappa_1 x_1})$, the complete integral (1.19) is estimated in the limit $\epsilon \to 0$ as

$$q = e^{i\omega_0 t} \int_{S_+} \frac{(-2\pi i) F(\mathbf{k})}{\partial B(\omega_0, \mathbf{k})/\partial k_1} e^{-ik_1 x_1} dk_2 dk_3. \tag{1.27}$$

Here, the outer integration with respect to k_2 and k_3 is confined to a surface S_+ defined as the set of all real wavenumber vectors $\mathbf{k}$ where the two conditions (1.25) are satisfied. Explaining this notation, we may remark that S is used for complete wavenumber surface $\omega(\mathbf{k}) = \omega_0$ while the part of S on which $\partial\omega/\partial k_1$ is positive is called S_+. The method of Lighthill (1960, 1978) concludes by estimating the integral (1.27) over S_+ by classical stationary-phase principles (involving contributions from just those points on S_+ where k_1 is stationary as a function of k_2 and k_3) as will be briefly recalled in §1.10 below – and then applied to a problem (§1.11) which was strongly interesting Fritz Ursell just before his appointment to the Beyer Chair (Ursell 1960).

1.6 Reconsideration of what poles lie in the strip

In the meantime I reconsider the issue of whether the enumeration (1.23) of poles in a narrow strip beneath the real k_1-axis includes any poles other than those just described; namely, poles which are on the real k_1-axis for $\epsilon = 0$ but which satisfy the inequality (1.24) so that they lie in the strip for $\epsilon > 0$. Clearly, other poles (if they exist) must be values of k_1 which even for $\epsilon = 0$ have

$$-\kappa_1 < \operatorname{Im} k_1 < 0. \tag{1.28}$$

Simultaneously they must satisfy

$$\omega(\mathbf{k}) = \omega_0; \text{ that is, } B(\omega_0, \mathbf{k}) = 0 \tag{1.29}$$

for ω_0 real (and, of course, for real k_2 and k_3); while the inequality (1.28) shows them to lie, not on the real wavenumber surface, but (for modest-sized κ_1) on a closely neighbouring part of the whole complex wavenumber surface S which (1.29) defines.

Using this remark as a clue, we are guided to look for the remaining poles on such straight lines

$$k_2 = \text{constant}, \ k_3 = \text{constant} \ (k_2, k_3 \text{ real}) \tag{1.30}$$

as may be described as "just missing" the real wavenumber surface S. Figure 1.2 illustrates two such lines; as the equation (1.29) for S shows, they must be lines on which $B(\omega_0, k_1, k_2, k_3)$ as a function of k_1 has *either* a minimum that is positive but very small or a maximum that is a very small negative number. In either case, evidently, a modest pure-imaginary change in k_1 from its value at the stationary point can give $B = 0$. In general this procedure will yield two poles (complex conjugates) depending on whether the change in $\operatorname{Im} k_1$ is positive or negative, but only the pole with $\operatorname{Im} k_1$ negative satisfies the condition (1.28), thus making an additional asymptotic contribution to the inner integral: a contribution first analysed by Lighthill (1990) in a detailed study of which a summary version is now given.

Figure 1.2 suggests that this contribution is associated with the edge C of the open surface S_+ defined by (1.25). This curve C which forms the edge of S_+ must be a locus on which

$$\partial \omega / \partial k_1 = 0 \tag{1.31}$$

so that the normal to S is perpendicular to the k_1-axis. Near a point P on C there can be lines (1.30) which "just miss" intersecting the real wavenumber surface S, yet for which there may in the strip (1.28) be a pole

$$k_1 = k_1^{\mathrm{P}} - im, \quad \text{with} \quad 0 < m < \kappa_1. \tag{1.32}$$

We may describe such a pole (1.32) as lying on a certain complex part S_{-i} of the wavenumber surface; specifically, a part extending from the edge C of S_+ in the negative pure-imaginary k_1-direction.

We proceed now to compare the asymptotic contributions from S_+ and from S_{-i} to our integrals. Intriguingly, we find these contributions to display an elegant mutual complementarity.

1.7   Formulating the contribution from S_{-i} to our integrals

We first review the local geometry of S (see Figure 1.2) near a point P on that curve C which is defined by equation (1.31) making the normal to S perpendicular to the k_1-axis. Near P we display in Figure 1.2 the intersection of S by a plane parallel to the k_1-direction and including the vector n, which we specify as the normal to S *towards which* the said intersection is convex.

In this plane, if we consider a small vector displacement from P whose component along n is h, we are able to approximate B at points on n by

$$(\partial B/\partial n)^P h \tag{1.33}$$

and at general points by

$$(\partial B/\partial n)^P [h + \frac{1}{2} R^{-1}(k_1 - k_1^P)^2], \tag{1.34}$$

where R is the radius of curvature at P of the plane section of S shown in Figure 1.2. This approximate expression (1.34) for B tells us firstly that a pole $B = 0$ takes the form (1.32) with, approximately,

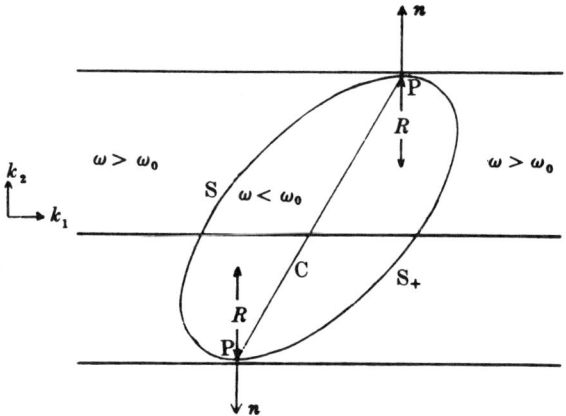

Figure 1.2: A particular wavenumber surface S on which $\omega = \omega_0$; a case when $\omega < \omega_0$ inside S and $\omega > \omega_0$ outside S is shown. That part S_+ of S on which $\partial \omega/\partial k_1 > 0$ is bounded by the curve C on which $\partial \omega/\partial k_1 = 0$. Three horizontal lines, defined as in the equations (1.30), are shown. The middle line intersects S at two real points, out of which only one is an admissible pole because it is on S_+. Each of the other two lines just misses intersecting S at any real point; but it intersects S at a pair of complex conjugate points out of which the point (1.32) is an admissible pole. Here, at the point P on C, the vector n is defined as the normal to S towards which its intersection by a plane including n and the k_1-direction is convex (with radius of curvature R).

$$h = \frac{1}{2}R^{-1}m^2. \tag{1.35}$$

(For modest-sized κ_1, then, h is indeed small.) Secondly, it tells us that (in the limit $\epsilon \to 0$) the contribution to the inner integral (1.20) in the form $(-2\pi i)$ times the residue at that pole is approximately

$$-2\pi i \, \frac{F\left(k_j^P\right) e^{-ik_1 x_1}}{(\partial B/\partial n)^P R^{-1}\left(k_1 - k_1^P\right)} \tag{1.36}$$

(the denominator being the derivative of expression (1.34) with respect to k_1); so that the asymptotic contribution from S_{-i} to the complete integral (1.19) is

$$-2\pi i e^{i\omega_0 t} \int_{S_{-i}} \frac{F\left(k_j^P\right) e^{-ik_1 x_1}}{(\partial B/\partial n)^P R^{-1}(k_1 - k_1^P)} \, dk_2 \, dk_3. \tag{1.37}$$

Next we display in Figure 1.3 a projection of the curve C onto a plane $k_1 = $ constant. In this plane, a neighbouring area element $dk_2 \, dk_3$ can be written $ds \, dh$ where ds is an element of arc length along the projected curve shown. But on S_{-i} (defined locally by expression (1.34) being equal to zero) we have

$$dh = d[-\frac{1}{2}R^{-1}(k_1 - k_1^P)^2] = -R^{-1}(k_1 - k_1^P)dk_1 \tag{1.38}$$

so that in the integral (1.37) the quotient

$$\frac{dk_2 \, dk_3}{R^{-1}(k_1 - k_1^P)} \tag{1.39}$$

takes a value

$$\frac{ds \, dh}{R^{-1}(k_1 - k_1^P)} = ds(-dk_1) \tag{1.40}$$

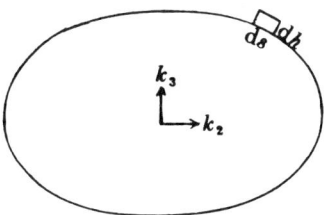

Figure 1.3: In a plane $k_1 = $ constant, where ds is an element of length along the projection of the curve C onto that plane, a nearby area element $dk_2 \, dk_3$ may be written $ds \, dh$, where h is distance along the normal direction **n**.

with h increasing from zero, which implies k_1 proceeding from k_1 to $k_1^P - i\kappa_1$. Alternatively (and more conveniently) we can replace (1.40) by

$$ds\, dk_1 \tag{1.41}$$

with k_1 going the other way from $k_1^P - i\kappa_1$ to k_1^P.

Finally, the contribution (1.37) from S_{-i} to the integral expression can be rewritten

$$-2\pi i e^{i\omega_0 t} \int_C ds \int_{k_1^P - i\kappa_1}^{k_1^P} \frac{F(k_j^P)}{(\partial B/\partial n)^P} e^{-ik_1 x_1}\, dk_1. \tag{1.42}$$

In §1.8 we compare this contribution from S_{-i} to a similar local form of the contribution (1.27) from S_+.

1.8   Behaviour near C of the contribution from S_+ to our integrals

On S_+, indeed, within a similarly small distance (less than κ_1) of a point P on its edge C, we have h negative and

$$dk_2\, dk_3 = ds\, d(-h) = ds\, d[\tfrac{1}{2} R^{-1}(k_1 - k_1^P)^2] \tag{1.43}$$

so that

$$\frac{dk_2\, dk_3}{\partial B(\omega_0, \mathbf{k})/\partial k_1} = \frac{ds\, R^{-1}(k_1 - k_1^P)\, dk_1}{(\partial B/\partial n)^P R^{-1}(k_1 - k_1^P)} = \frac{ds\, dk_1}{(\partial B/\partial n)^P}. \tag{1.44}$$

The local behaviour of the integral (1.27) over S_+ is therefore as

$$-2\pi i e^{i\omega_0 t} \int_C ds \int_{k_1^P}^{k_1^P + \kappa_1} \frac{F(\mathbf{k})}{(\partial B/\partial n)^P} e^{-ik_1 x_1}\, dk_1. \tag{1.45}$$

We can now investigate the overall modification to the asymptotic theory (Lighthill 1960, 1978) given in §1.5 which results from careful consideration of the neighbourhood of the edge C of S_+. Our preliminary insight (§1.6) was that the asymptotic form (1.27) needs to be augmented so as to become

$$-2\pi i e^{i\omega_0 t} \left(\int_{S_+} + \int_{S_{-i}} \right) \frac{F(\mathbf{k}) e^{-ik_1 x_1}}{\partial B(\omega_0, \mathbf{k})/\partial k_1}\, dk_2\, dk_3. \tag{1.46}$$

In §§1.7 and 1.8, furthermore, we have shown local behaviours (1.42) and (1.45) for this pair of integrals (1.46), which may be combined into an overall local behaviour like

$$-2\pi i e^{i\omega_0 t} \int_C ds \left(\int_{k_1^P - i\kappa_1}^{k_1^P} + \int_{k_1^P}^{k_1^P + \kappa_1} \right) \frac{F(k_j^P)}{(\partial B/\partial n)^P} e^{-ik_1 x_1}\, dk_1. \tag{1.47}$$

The importance of this local behaviour (1.47) around the edge C of S_+ was shown by Lighthill (1990) to lie in the absence of any discontinuity. Thus, in the combined inner integral, the integrand exhibits no discontinuity at $k_1 = k_1^P$, even though (Figure 1.4) the path of integration turns a corner at that point.

By contrast, either inner integral in (1.47) by itself has effectively a discontinuity at the point k_1^P (a representative point on the edge C) and this discontinuity gives it an $O(x_1^{-1})$ asymptotic behaviour

$$\pm \frac{F(k_j^P)}{(\partial B/\partial n)^P} \frac{e^{-ik_1^P x_1}}{(-i\, x_1)} \tag{1.48}$$

(with the $+$ sign for the first, and the $-$ sign for the second, inner integral). These asymptotic results follow from standard asymptotic theory for Fourier and Laplace transforms, but the key fact that they cancel out in the combined integral (1.47) results explicitly from the absence of any discontinuity in the inner integral as taken along the right-angled path in Figure 1.4.

To sum up this section, the integral (1.27) over S_+ has near the edge C of S_+ a local behaviour as in (1.45), where the inner integral takes the form of a Fourier integral of a function of k_1 with a discontinuity at k_1^P. This discontinuity would generate its own $O(x_1^{-1})$ asymptotic behaviour if the integral (1.42) over S_{-i} were not also present to remove the discontinuity.

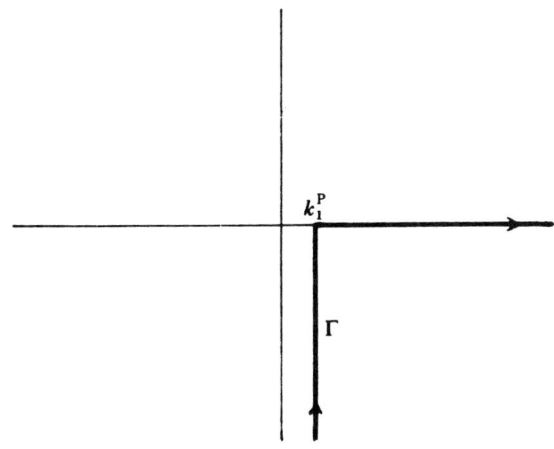

Figure 1.4: The inner integral in expression (1.47) takes the form of an integral along the path Γ, having a right-angled corner at $k_1 = k_1^P$; and its asymptotic behaviour includes no term of order x_1^{-1} corresponding to k_1^P because k_1^P is not a point of discontinuity of the integrand.

1.9   Essential role of the integral over S_{-i} as a continuation of that over S_+

The fact that integration over the surface S_+ (a real part of S, specified by the inequality $\partial\omega/\partial k_1 > 0$) is continued over S_{-i} (a complex part of S, extending from the edge C of S_+ in the negative pure-imaginary direction) is now seen, therefore, to be essential for ensuring that no $O(x_1^{-1})$ asymptotic contribution arises from the neighbourhood of C. It follows that the only asymptotic contribution of this order can be correctly calculated by stationary-phase methods applied directly to the integral over S_+; this calculation is comprehensively set out in §§4.9 – 4.12 of Lighthill (1978), while some specially important aspects are briefly recalled in §§1.10 and 1.11 below.

I am grateful to Professor V.A. Borovikov of the USSR Academy of Sciences Institute for Problems of Mechanics, Moscow, who wrote to me demonstrating by means of two simple examples that the integral taken just over S_+ can give an asymptotically incorrect result. His letter prompted me in due course to work out my emendation as given above (and see Lighthill (1990) for fuller details) to the original proof by Lighthill (1960, 1978) of the asymptotic behaviour of anisotropic wave systems stimulated by oscillating sources.

1.10   Some important asymptotic results in two or three dimensions

Having established after reconsideration that an application of stationary-phase principles to the integral (1.27) over S_+ does yield correct asymptotic results, I briefly recall some of the main conclusions which can be derived in this way. They are given first in two dimensions – when S is a curve in the (k_1, k_2) plane – and then in three dimensions with S a surface in the wavenumber space (1.5); however, for systems where the curvature of S may be zero either globally or locally, the reader is referred to §§4.10 and 4.11 (respectively) of Lighthill (1978).

In the case of two-dimensional propagation the asymptotic expression (1.27) with the k_3-integration omitted is estimated in terms of contributions from any point of stationary phase on the curve S_+ defined by the conditions (1.25). The phase $(-k_1 x_1)$ is stationary on S_+ at a point

$$(k_1^{(0)}, k_2^{(0)}) \quad \text{where} \quad dk_1/dk_2 = 0. \tag{1.49}$$

Near such a stationary point S_+ is approximated by the equation

$$k_1 = k_1^{(0)} + \frac{1}{2}(d^2 k_1/dk_2^2)^{(0)}(k_2 - k_2^{(0)})^2 \tag{1.50}$$

and stationary-phase principles specify the contribution from such a stationary point to the asymptotic form of (1.27) as

$$-2\pi i e^{i\omega_0 t} \frac{F(\mathbf{k}^{(0)}) e^{-ik_1^{(0)} x_1}}{[\partial B(\omega_0, \mathbf{k})/\partial k_1]^{(0)}} \frac{\pi^{1/2} e^{-\frac{1}{4}\pi i \operatorname{sgn}(d^2 k_1/dk_2^2)^{(0)}}}{[\frac{1}{2}|d^2 k_1/dk_2^2|^{(0)} x_1]^{\frac{1}{2}}}; \tag{1.51}$$

being derived from (1.27) by (i) replacing k_1 in the exponential by just the terms (1.50), while (ii) replacing k_1 by $k_1^{(0)}$ elsewhere, and (iii) completing the calculation as a Gaussian integral (this is where the $\pi^{1/2}$ factor emerges).

For convenience, we express this asymptotic result (1.51) in a form invariant under rotation of axes, that can then be applied to wave generation problems without the need for any preliminary change of axes. The problem being studied (see §1.5) is to estimate q on an arbitrary line ℓ stretching radially out from the origin. The condition (1.49) which defines the stationary point on S_+ when axes are chosen with ℓ as the positive x_1-axis means that a direction normal to S_+ is in the direction of ℓ. Furthermore, the definition (1.25) of S_+ specifies this normal direction to the surface $\omega(\mathbf{k}) - \omega_0$ as the normal in the direction ω increasing. Thus, the point $\mathbf{k}^{(0)}$ can in general axes be defined by the condition that a unit vector $\mathbf{n}$ directed along ℓ must coincide with

the normal $\mathbf{n}$ to S at $\mathbf{k}^{(0)}$ in the direction ω increasing.   (1.52)

Also, the quantity $|d^2 k_1/dk_2^2|^{(0)}$ becomes in general axes the curvature $\kappa^{(0)}$ of S at the point $\mathbf{k}^{(0)}$, while $\mathrm{sgn}(d^2 k_1/dk_2^2)^{(0)}$ is -1 or $+1$ according as S at $\mathbf{k}^{(0)}$ is convex or concave towards ℓ; and, finally, $k_1^{(0)} x_1$ becomes $\mathbf{k}^{(0)} \cdot \mathbf{x}$.

We conclude that, at any position $\mathbf{x} = x\mathbf{n}$ where $\mathbf{n}$ is a unit vector and x is large, the asymptotic form of q can be expressed in terms of the point $\mathbf{k}^{(0)}$ on S for which $\mathbf{n}$ is the normal $(1.52)^2$ as

$$q \sim -2\pi i e^{i\omega_0 t \pm \frac{1}{4}\pi i} \frac{F(\mathbf{k}^{(0)}) e^{-i x \mathbf{k}^{(0)} \cdot \mathbf{n}}}{(\partial B/\partial n)^{(0)}} \left(\frac{2\pi}{\kappa^{(0)} x} \right)^{\frac{1}{2}},   (1.53)$$

with the sign in front of $\frac{1}{4}\pi i$ being $+$ or $-$ according as S is convex or concave to $\mathbf{n}$ at $\mathbf{k}^{(0)}$. The appearance in (1.53) of $\kappa^{(0)}$, the curvature of S at $\mathbf{k}^{(0)}$, has a simple physical interpretation from the fact (Figure 1.5) that two directions normal to S at neighbouring points a distance ds apart include an angle $\kappa^{(0)} ds$, so that the distance between them becomes $\kappa^{(0)} x ds$ for large x. Equation (1.53) is consistent, therefore, with the idea of a uniform energy flux between adjacent rays.

Here, pursuing this point a little farther in preference to any geometrically elegant analysis – as, for example, in Lighthill (1978), Figure 90 – of the corresponding phase information, we write down the total wave energy flux generated by the sources; in other words, their wavemaking power P. For a general wave system where the energy density W for waves of wavenumber $\mathbf{k}$ can be written

$$W = W_0(\mathbf{k}) \bar{q}^2   (1.54)$$

---

[2] If there are two or more such points the asymptotic form of q is the sum of the terms (1.53) corresponding to each

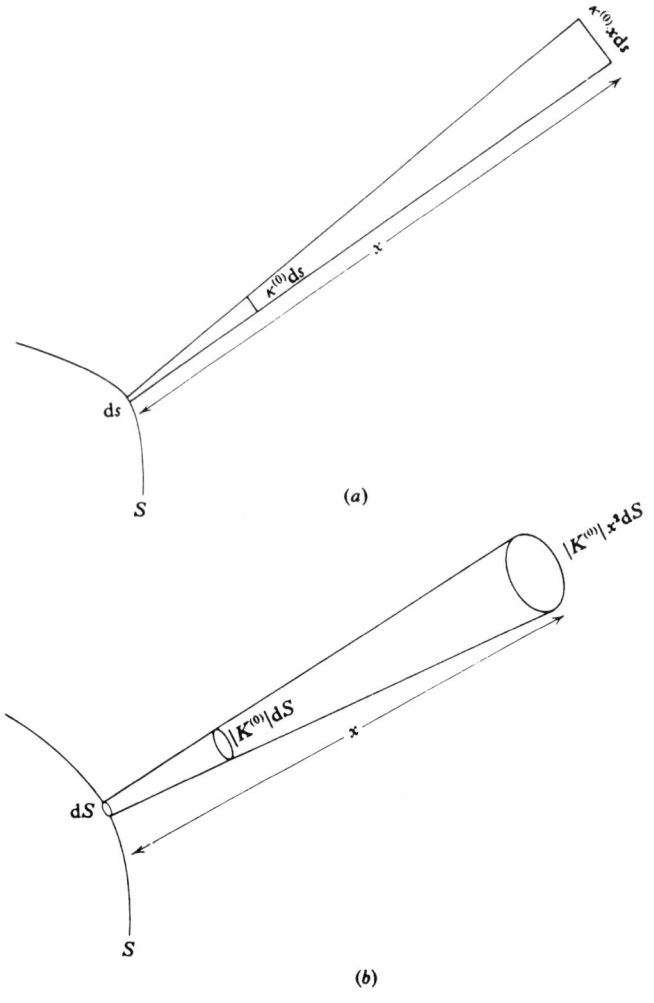

Figure 1.5: Spreading of the energy that travels between neighbouring directions normal to S. (a) Two-dimensional case: two normal directions include an angle $\kappa^{(0)}ds$, where ds is a length-element of the wavenumber curve S and $\kappa^{(0)}$ is its curvature; the separation between them is $\kappa^{(0)}xds$ at a large distance x. (b) Three-dimensional case: a cone of normals to the wavenumber surface S includes a solid angle $\mid K^{(0)} \mid dS$, where dS is an area-element of S and $K^{(0)}$ is its Gaussian curvature; the cross-sectional area of the cone is therefore $\mid K^{(0)} \mid x^2dS$ at a large distance x.

in terms of the mean square value of q, equation (1.53) gives the directional distribution of wave energy density as

$$W = \frac{W_0(\mathbf{k}^{(0)})|F(\mathbf{k}^{(0)})|^2}{[(\partial B/\partial n)^{(0)}]^2} \frac{4\pi^3}{\kappa^{(0)}x}. \tag{1.55}$$

The corresponding energy flux is $W\mathbf{U}$ where $\mathbf{U}$ is the group velocity vector, with a magnitude $U = \partial\omega/\partial n$ that takes the value

$$U = -\frac{\partial B/\partial n}{\partial B/\partial \omega} \tag{1.56}$$

for the dispersion relationship (1.7). The power output P can now be obtained from (1.55) by multiplying the magnitude of the energy flux by the distance $\kappa^{(0)}x ds$ (see Figure 1.5) between adjacent rays associated with a length-element ds of S and integrating around S to give

$$P = 4\pi^3 \int_S \frac{W_0(\mathbf{k})|F(\mathbf{k})|^2}{|(\partial B/\partial \omega)(\partial B/\partial n)|} ds; \tag{1.57}$$

here, the superscript (0) used to pick out a point on S with normal in a particular direction may obviously be dropped.

The simple expression (1.57) for wavemaking power, in the form of the spectrum $|F(\mathbf{k})|^2$ of the source strength multiplied by a factor dependent on the wave system and integrated around the curve S on which $B(\omega_0, \mathbf{k}) = 0$, proves invaluable in a wide range of contexts. Just one of these will be indicated in §1.11 below.

In three-dimensional problems, with S as the wavenumber surface, the asymptotic result (1.53) needs to be replaced by

$$q \sim -2\pi i e^{i\omega_0 t \pm \frac{1}{4}\pi i \pm \frac{1}{4}\pi i} \frac{F(\mathbf{k}^{(0)})e^{-ix\mathbf{k}^{(0)}\cdot\mathbf{n}}}{(\partial B/\partial n)^{(0)}} \frac{2\pi}{|K^{(0)}|^{1/2}x} \tag{1.58}$$

where $K^{(0)}$ is the Gaussian curvature of S (product of the two principal curvatures) at the point $\mathbf{k}^{(0)}$ and each $\pm$ sign is $+$ or $-$ according as a principal curvature of the surface is convex or concave to n. Expression (1.58) is derived from an analysis extremely similar to that just given; in which, however, the last fraction in expression (1.51) is replaced by the product of two such fractions – one for each of the principal directions of curvature. Geometrically, a cone of normals to S from an area-element dS includes a solid angle $|K^{(0)}|dS$, so that (see Figure 1.5) the cross-sectional area of the cone is $|K^{(0)}|x^2 dS$ at a large distance x. It follows once again that in the expression for total wavemaking power P the curvatures cancel out, to give P in the valuable form of a surface integral

$$P = 8\pi^4 \int_S \frac{W_0(\mathbf{k})|F(\mathbf{k})^2}{|(\partial B/\partial \omega)(\partial B/\partial n)|} dS \tag{1.59}$$

over the wavenumber surface S defined by the equation $B(\omega_0, \mathbf{k}) = 0$.

Against the general background of this paper, the significance of the $O(x^{-1})$ asymptotic result (1.58) is that an asymptotic expression of just the same order would arise (§1.8) from any discontinuity at C of the integral over S_+ alone. Fortunately, however, contamination by such an additional $O(x^{-1})$ term is not really present, because the integral over S_+ is necessarily supplemented by one over S_{-i} in such a way that any discontinuity is eliminated.

1.11   Wave generation by travelling forcing effects

It is also most fortunate that the generation of waves by travelling forcing effects of fixed frequency ω_0 can be calculated by a completely straightforward extension of the theory for stationary sources. We shall see moreover that (i) because source motion makes even an isotropic system (such as surface gravity waves) effectively anisotropic, the theory of §§1.2–1.10 for anisotropic wave systems is necessarily appropriate; while (ii) the case $\omega_0 = 0$ (generation of waves by a travelling steady source – like a ship moving uniformly) now becomes interesting and by no means trivial.

I shall take the source velocity as $(-\mathbf{V})$ relative to undisturbed fluid. This arrangement is completely equivalent to the case of a source at rest generating waves in a uniform stream of fluid of velocity $\mathbf{V}$. Both these equivalent problems are important; and, furthermore, we aim in both problems to calculate the waves generated in the same frame of reference; that is, relative to the source.

The distribution of source strength for a forcing effect of frequency ω_0 travelling at velocity $(-\mathbf{V})$ may be written

$$e^{i\omega_0 t} f(\mathbf{x} + \mathbf{V}t), \tag{1.60}$$

and when $f(\mathbf{x})$ is given by a Fourier integral (1.4) this takes the form

$$\int F(\mathbf{k}) e^{i[\omega_0 t - \mathbf{k}\cdot(\mathbf{x}+\mathbf{V}t)]} d\mathbf{k}; \tag{1.61}$$

in which components of wavenumber $\mathbf{k}$ have a Doppler-shifted frequency

$$\omega_0 - \mathbf{k} \cdot \mathbf{V} = \omega_\mathrm{r}, \tag{1.62}$$

also known as the *relative* frequency. It immediately follows (as in §1.3), therefore, that the wave motions corresponding to the forcing effect (1.61) are given by the integral expression

$$q = \int \frac{F(\mathbf{k}) e^{i[\omega_0 t - \mathbf{k}\cdot(\mathbf{x}+\mathbf{V}t)]}}{B(\omega_0 - \mathbf{k}\cdot\mathbf{V}, \mathbf{k})} \, d\mathbf{k}; \tag{1.63}$$

and so the entire previous theory can be directly applied with

(i) $\mathbf{x}$ replaced by $\mathbf{x} + \mathbf{V}t = \mathbf{X}$ (position relative to the source);

and

(ii) $B(\omega_0, \mathbf{k})$ replaced by $B(\omega_0 - \mathbf{k} \cdot \mathbf{V}, \mathbf{k}) = B_{\mathrm{r}}(\omega_0, \mathbf{k})$.

In particular, S becomes the Doppler-shifted wavenumber surface (or curve in two dimensions)

$$B_{\mathrm{r}}(\omega_0, \mathbf{k}) = 0. \tag{1.64}$$

These are beautifully simple modifications to the theory as given earlier for stationary sources. I conclude by illustrating them from the theory of ship waves.

For two-dimensional propagation, the result (1.53) adapted to the case of travelling forces effects becomes

$$q \sim -2\pi i e^{-i\omega_0 t \pm \frac{1}{4}\pi i} \frac{F(\mathbf{k}^{(0)}) e^{iX\mathbf{k}^{(0)} \cdot \mathbf{n}}}{(\partial B_{\mathrm{r}}/\partial n)^{(0)}} \frac{2\pi}{\kappa^{(0)} X} \tag{1.65}$$

and the wavemaking power (1.57) becomes

$$P = 4\pi^3 \int_S \frac{W_0(\mathbf{k})|F(\mathbf{k})|^2}{|(\partial B_{\mathrm{r}}/\partial \omega)(\partial B_{\mathrm{r}}/\partial n)|} \, ds. \tag{1.66}$$

When $B(\omega, \mathbf{k})$ takes the form (1.15) appropriate to gravity waves on deep water, and when $\mathbf{V} = (V, 0)$ corresponding to a ship moving at speed V in the negative x_1-direction, then

$$B_{\mathrm{r}} = B(\omega_0 - k_1 V, \mathbf{k}) = \rho(\omega_0 - k_1 V)^2 (k_1^2 + k_2^2)^{-1/2} - \rho g. \tag{1.67}$$

In particular, for $\omega_0 = 0$ the curve S defined by equation (1.64) has two branches as shown in Figure 1.6, where the arrows indicate normals $\mathbf{n}$ in the direction ω_0 increasing. These vectors $\mathbf{n}$ correspond to the directions behind the ship where the theory predicts the presence of waves; directions which for the pattern of ship waves fill, as expected, the familiar Kelvin wedge of semi-angle $\sin^{-1}\left(\frac{1}{3}\right) = 19\frac{1}{2}^{\circ}$ that has been analyzed so penetratingly by Fritz Ursell himself (Ursell 1960).

Next, within the expression (1.66) for wavemaking power, the factor $W_0(\mathbf{k})$, specified in §1.10 as that by which the mean square of q (here, the surface elevation) needs to be multiplied to give the wave energy density, assumes a constant value ρg for gravity waves (since their potential energy per unit horizontal area is $\frac{1}{2}\rho g q^2$ and, in the mean, the potential and kinetic energies must be equal). Substituting this value for $W_0(\mathbf{k})$, and expression (1.67) with $\omega_0 = 0$ for B_{r}, we obtain after some reduction

$$P = \frac{2\pi^3 V}{\rho g} \int_S \frac{k_1^2 + k_2^2}{k_1^2 + 2k_2^2} \mid k_1 F(\mathbf{k}) \mid^2 \mid dk_2 \mid . \tag{1.68}$$

The integral here is quite a simple integral along the wavenumber curve of the quantity $|k_1 F(\mathbf{k})|^2$ which may be called the spectrum of $\partial f / \partial x_1$, modified by a preceding factor which varies only between 1 and $\frac{1}{2}$. The

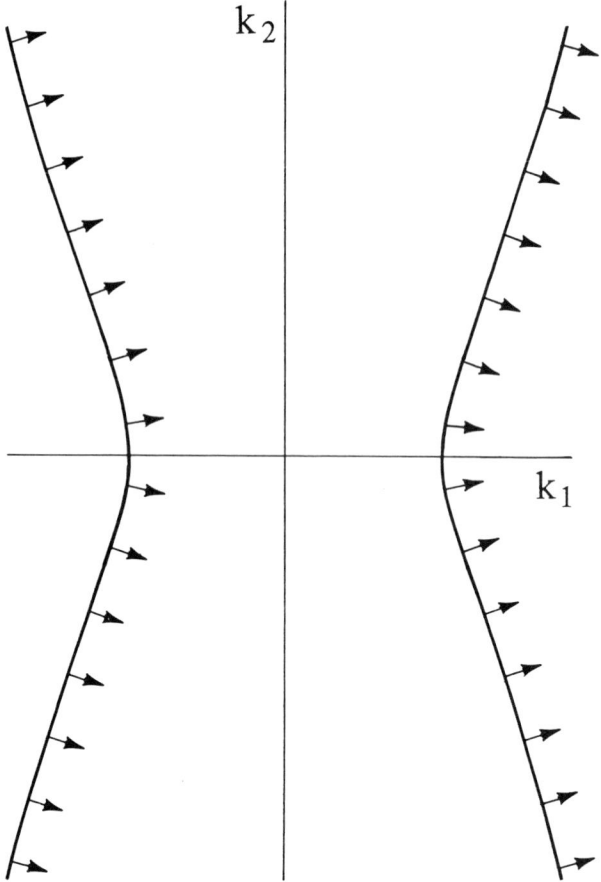

Figure 1.6: The wavenumber curve S for waves generated by a ship moving at steady velocity V in the negative x_1-direction. Here, S is the locus (1.64) for $\omega_0 = 0$, with B_r given by (1.67). Waves are found in directions specified by the arrows, representing normals to the curve in the direction ω_0 increasing. The distance from the origin to the nearest points on S is gV^{-2}.

value of the wavemaking resistance corresponding to expression (1.68) is simply $V^{-1}P$.

These results can be used just as they stand to calculate the wavemaking resistance of a steadily moving hovercraft, which applies to the sea surface a pattern of excess pressure distribution (1.12) in the form

$$\eta = f(\mathbf{x} + \mathbf{V}t), \tag{1.69}$$

of which the Fourier transform $F(\mathbf{k})$ appears directly in equation (1.68). But the results, as Lighthill (1978) showed, are also relevant to the determination of wavemaking resistance for conventional ships.

The distribution of η equivalent to an immersed ship may often by estimated to sufficient accuracy by writing the velocity potential ϕ as

$$\phi = \phi_h + \phi_w, \tag{1.70}$$

where ϕ_h represents an irrotational flow that would be generated in unbounded fluid by the hull's motion and ϕ_w is the velocity potential of the waves. Then the boundary value $\eta = \eta_w$ corresponding to the waves must satisfy the condition

$$\eta_w = -\eta_h \tag{1.71}$$

if the complete solution (1.70) is to satisfy the surface condition $\eta = 0$.

Those concerned with determining the frictional resistance of a hull form often use a "double model", obtained by adding on to a model of the immersed part of the hull an inverted but otherwise identical shape which is its mirror image in the water surface. Experiments on such a double model in a fully immersed condition may be used, moreover, to determine not only frictional drag but also η_h, the distribution of excess pressure on its horizontal plane of symmetry. (As an alternative, this distribution is also straightforward to compute.) Smoothly extrapolated values must be used in that part of the plane of symmetry that actually intersects the hull. Then the waves are generated by a source η_w given by equation (1.71) in a form (1.69),[3] and the wavemaking power (1.68) is readily deduced.

These facts help us to understand the effectiveness of certain types of modification of hull forms in deferring to relatively higher velocities any steep rise in the wavemaking resistance $V^{-1}P$ as a function of ship velocity V. Figure 1.6 shows that the lowest magnitude of the wavenumber on the curve S is gV^{-2}. Accordingly, the expression (1.68) begins to be substantial when V rises to such a value that the gradient $\partial f/\partial x_1$

---

[3]We may note that η_w is symmetrical about the ship's own vertical plane of symmetry, so that ϕ_w has zero normal derivative on that plane; which means that ϕ_w makes little disturbance (none at all on the "thin ship" approximation) to the satisfaction by ϕ_h of the boundary condition on the hull's surface

of pressure in the central plane of such a double model has significant spectral components with wavenumbers as high as gV^{-2}. On the other hand, such a steep rise may be deferred to rather higher velocities by choice of hull forms where spectral components of $\partial f / \partial x_1$ with such relatively high wavenumber are "ironed out".

Near the bow of a conventional ship form, Figure 1.7 shows that the form of the computed excess pressure η_h includes large gradients varying rather steeply in the region just ahead of the double model's bow where the excess pressure rises to its maximum value. These may be smoothed out by introducing an element of bulbous shape below the surface in this region. The double model then includes two bulbous elements. By themselves they generate a pressure reduction on the central plane due to acceleration of the water relative to the model as it travels between them (Venturi effect). Careful design can make this distribution of negative excess pressure, due to the sub-surface bulbous element at the bow, cancel out enough of the distribution of positive excess pressure due to the main hull so that the spectral components of $\partial f / \partial x_1$ with higher wavenumber are greatly reduced. In classic experiments Inui (1962) first demonstrated that such bulbous bows can substantially defer the steep rise of wave-making resistance with speed – and their subsequent utilisation by naval architects has improved significantly the economics of shipping.

I have concluded this lecture with a very brief look at one particular application (to the objective of understanding the effectiveness of techniques for reducing wave-making resistance) for the general theory of anisotropic wave asymptotics. Important practical applications of the theory are, however, extremely widespread (see §§4.9 − 4.12 of Lighthill (1978)); it is reassuring therefore, that the present review of its fundamentals has given it "a clean bill of health".

1.12   References

Inui, T. 1962 Wave-making resistance of ships. *Trans. Soc. Nav. Arch. Mar. Eng.* **70**, 282–353.

Lighthill, M.J. 1960 Studies on magnetohyrodynamic waves and other anisotropic wave motions. *Phil. Trans. Roy. Soc. A* **252**, 397–430.

Lighthill, J. 1978 *Waves in fluids.* Cambridge: University Press.

Lighthill, J. 1990 Emendations to a proof in the general three-dimensional theory of oscillating sources of waves. *Proc. Roy. Soc. A* **427**, 31–42.

Ursell, F. 1960 On Kelvin's ship-wave pattern. *J. Fluid Mech.* **8**, 418–431.

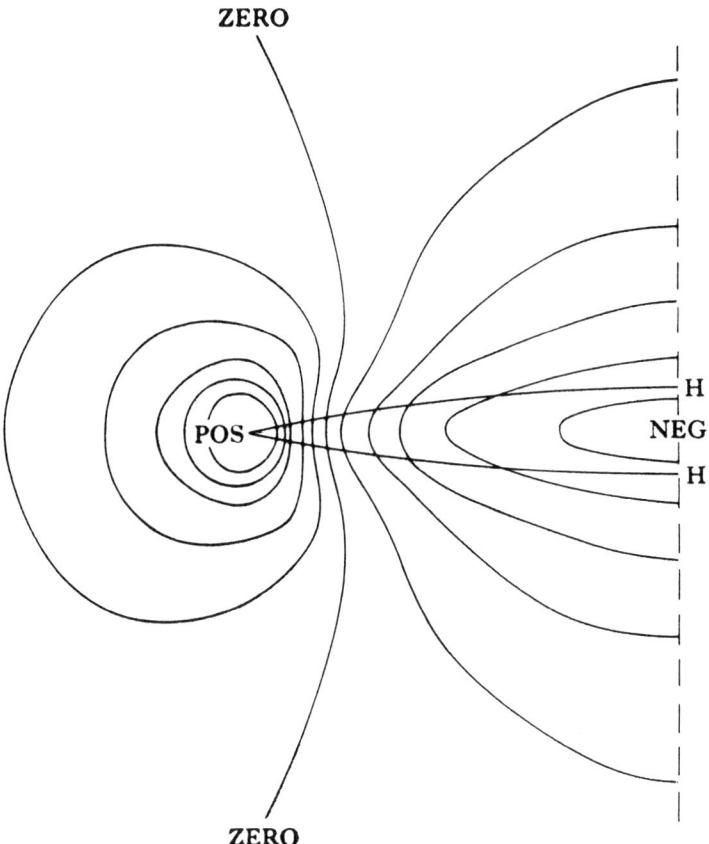

Figure 1.7: The distribution η_h of excess pressure in the plane of symmetry of a fully immersed double model of a ship hull H. These contours of equally spaced constant values of η_h are derived from a linear-theory computation for a parabolic ship with vertical sides and with draught-length ratio 0.25. The positive pressure peak POS near the bow is characteristic, however, of all sharp-bowed ship shapes. (This calculation, made to exhibit that intense localised pressure peak, pays no special attention to the stern and in fact takes the ship and the flow as symmetrical about the broken line.)

2

Uniform asymptotics of oscillating integrals: applications in chemical physics

J.N.L. Connor, P.R. Curtis and R.A.W. Young
University of Manchester

Fritz Ursell has made fundamental contributions to the uniform asymptotics of oscillating integrals. His ideas can be applied to many problems in chemical physics involving the dynamics of atoms and molecules. In this chapter, we describe two new applications of uniform asymptotic techniques, which are closely related to his papers of 1972 and 1980. The first application concerns Franck-Condon integrals and the uniform swallowtail approximation, whilst the second applies the uniform hyperbolic umbilic approximation to an oscillating integral suggested by the theory of atom-surface collisions.

2.1   Introduction

The asymptotic evaluation of integrals is an important subject, with numerous applications to high frequency scattering phenomena. In particular, the asymptotic problem of uniformly approximating oscillating integrals with several coalescing saddle points arises in many short wavelength theories.

The simplest case concerns one-dimensional integrals of the form

$$I(\alpha) = \int_{-\infty}^{\infty} g(t) \exp[if(\alpha; t)/\hbar]\, dt, \quad \hbar \to 0, \tag{2.1}$$

where $\alpha \equiv (\alpha_1, \alpha_2, \ldots)$ is a set of real parameters, and $g(t)$ and $f(\alpha; t)$ are analytic functions, with $f(\alpha; t)$ real for real t.

The saddle point (or steepest descent) method (Fedoriuk 1977; Olver 1974; Bleistein & Handelsman 1986; Wong 1989), shows that the main contribution to $I(\alpha)$ comes from the neighbourhood of the (real or complex) saddle points t_i of $f(\alpha; t)$, which obey

$$\frac{\partial f(\alpha; t)}{\partial t} = 0 \quad \text{for} \quad t = t_i(\alpha), \quad i = 1, 2, \ldots.$$

As α varies, the positions of the saddle points change; in particular they can come close together or coalesce.

For the case of two coalescing saddle points and a single parameter, a fundamental contribution was made in 1957 by Chester, Friedman and Ursell (1957) [see also Friedman (1959) and Ursell (1965)]. They introduced an exact local transformation whereby $f(\alpha; t)$ is mapped onto a cubic polynomial. After making this change of variable in the integral (2.1), and expanding the new pre-exponential factor in a novel way, Chester *et al.* (1957) showed that the uniform asymptotic approximation for the integral (2.1) could be written in terms of the regular Airy function and its first order derivative. The regular Airy function is the *canonical integral* for this problem.

Later work by Levinson (1961), Bleistein (1967), Ursell (1972) and Martin (1974) extended the case of two coalescing saddle points to that of an arbitrary number of coalescing saddle points. The language of elementary catastrophe theory (Poston & Stewart 1978; Connor 1976) provides a convenient way to characterize these uniform asymptotic expansions, as well as their generalisations to multidimensional integrals (Connor 1976). In particular, the uniform asymptotic approximation of one-dimensional integrals involves the cuspoid catastrophes, whereas for two-dimensional integrals, the umbilic catastrophes arise (Connor 1976).

In applications of this uniform asymptotic theory to concrete problems, only the case of two coalescing saddle points is easy to handle, which involves the regular Airy function (or fold canonical integral):

$$\text{Ai}(x) = \frac{1}{2\pi} \int_{-\infty}^{\infty} \exp[i(u^3/3 + xu)] \, du. \qquad (2.2)$$

However practical techniques have been developed at the University of Manchester to handle oscillating integrals which possess three (Connor & Farrelly 1981a, b; Connor & Curtis 1982; Connor, Curtis & Farrelly 1983) or four (Connor & Curtis 1982; Connor, Curtis & Farrelly 1983, 1984) coalescing saddle points [the methods actually apply to any number of coalescing saddle points in principle (Connor & Curtis 1984)].

The canonical integral for three coalescing saddle points is the Pearcey integral (or cusp canonical integral) (Connor 1973c; Pearcey 1946):

$$P(x, y) = \int_{-\infty}^{\infty} \exp[i(u^4 + xu^2 + yu)] \, du. \qquad (2.3)$$

For four coalescing saddle points, it is the swallowtail canonical integral (Connor & Curtis 1982; Connor, Curtis & Farrelly 1983, 1984; Connor 1974):

$$S(x, y, z) = \int_{-\infty}^{\infty} \exp[i(u^5 + xu^3 + yu^2 + zu)] \, du. \qquad (2.4)$$

The techniques developed at Manchester for the practical application of the uniform cuspoid asymptotic theory have recently been reviewed (Connor 1990). This review also describes briefly a large number of problems in chemical physics that involve oscillating integrals.

The purpose of the present chapter is to present two new applications of uniform asymptotic techniques. The integrals again come from problems in chemical physics. The first application concerns Franck-Condon integrals (Connor & Curtis, unpublished results) which are defined by

$$I_{FC}(\boldsymbol{\alpha}) = \int_{-\infty}^{\infty} \psi_1(x)\psi_2(x)\,dx, \qquad (2.5)$$

where $\psi_1(x)$ and $\psi_2(x)$ are solutions of the Schrödinger equation for two different potential energy functions. The integrand of equation (2.5) is not in the required form of equation (2.1), so the standard asymptotic techniques do not apply. This difficulty can be avoided by using suitable integral representations for $\psi_1(x)$ and $\psi_2(x)$, which allow I_{FC} to be written as a three-dimensional integral of the type

$$I_{FC}(\boldsymbol{\alpha}) = \int_{-\infty}^{\infty} \int_{-\infty}^{\infty} \int_{-\infty}^{\infty} g(x,s,t)\exp[if(\boldsymbol{\alpha};x,s,t)/\hbar]\,dx\,ds\,dt.$$

A catastrophe (or singularity) theory analysis of $f(\boldsymbol{\alpha};x,s,t)$ is then carried out, in order to determine the local polynomial form onto which $f(\boldsymbol{\alpha};x,s,t)$ can be mapped exactly. This form turns out to be a swallowtail plus a Morse 1-saddle for the example discussed in §§2.5–2.7. The Morse term in the three-dimensional integral can be integrated explicitly, leaving a one-dimensional integral to which the standard asymptotic techniques can then be applied.

The second application concerns the uniform hyperbolic umbilic approximation (Connor 1973a, b, 1976; Connor & Young, unpublished results). We have also developed practical methods that allow this appoximation to be applied to concrete problems. This is illustrated for a two-dimensional non-separable integral that occurs in the theory of atom-surface scattering.

This chapter is arranged in the following way. The lowest-order terms in the uniform cuspoid approximation are presented in §2.2. The analogous equations for the uniform hyperbolic umbilic approximation are reported in §2.3, whilst an application of this approximation to a non-separable two-dimensional integral is considered in §2.4. In §2.5, it is shown how the Franck-Condon integral for a linear-potential harmonic-oscillator system may be written as a three-dimensional integral. A catastrophe theory analysis of the exponent of this integral is discussed in §2.6. Uniform asymptotic results for the Franck-Condon integral are presented in §2.7. Our conclusions are in §2.8.

2.2   Uniform cuspoid approximation

This section presents the equations for the uniform cuspoid approximation. Suppose $f(\boldsymbol{\alpha};t)$ in equation (2.1) possesses n coalescing saddle points. The first step is to map locally $f(\boldsymbol{\alpha};t)$ onto the polynomial (Ursell 1972; Connor 1974):

$$f(\boldsymbol{\alpha};t) = u^{n+1} + \sum_{j=1}^{n-1} \zeta_j u^j + A, \tag{2.6}$$

where $u = u(\boldsymbol{\alpha};t)$, but the new parameters $\{\zeta_j\}$ and A depend just on $\boldsymbol{\alpha}$. Equation (2.6) defines an exact local one-to-one uniformly analytic transformation, provided the saddle points on either side correspond, i.e.

$$t = t_i(\boldsymbol{\alpha}) \longleftrightarrow u = u_i(\boldsymbol{\alpha}), \quad i = 1, 2, \dots, n.$$

After making the change of variable of equation (2.6), the integral $I(\boldsymbol{\alpha})$ becomes

$$I(\boldsymbol{\alpha}) = \exp(iA/\hbar) \int_{-\infty}^{\infty} g(t) \frac{dt}{du} \exp[i(u^{n+1} + \sum_{j=1}^{n-1} \zeta_j u^j)/\hbar] \, du. \tag{2.7}$$

Next, to obtain the lowest-order terms in the uniform approximation, the pre-exponential factor in equation (2.7) is expanded in the form (Ursell 1972; Connor 1974)

$$g(t)\frac{dt}{du} \approx \sum_{k=0}^{n-1} q_k u^k. \tag{2.8}$$

The uniform cuspoid approximation for $I(\boldsymbol{\alpha})$ is then given by

$$I(\boldsymbol{\alpha}) \sim \exp(iA/\hbar) \sum_{k=0}^{n-1} \hbar^{(k+1)/(n+1)} q_k \tag{2.9}$$

$$\times \quad U_k \left(\hbar^{-n/(n+1)} \zeta_1, \hbar^{-(n-1)/(n+1)} \zeta_2, \dots, \hbar^{-2/(n+1)} \zeta_{n-1} \right),$$

where

$$U_k(\zeta_1, \zeta_2, \dots, \zeta_{n-1}) \equiv \int_{-\infty}^{\infty} u^k \exp[i(u^{n+1} + \sum_{j=1}^{n-1} \zeta_j u^j)] \, du. \tag{2.10}$$

Note that for $n = 2, 3, 4$, the integrals $U_k(\zeta_1, \zeta_2, \dots, \zeta_{n-1})$ for $k = 0, 1, \dots,$ $n - 1$ are closely related to $\mathrm{Ai}(x)$, $P(x,y)$, $S(x,y,z)$ and their first order partial derivatives [see equations (2.2)–(2.4)].

An inspection of equations (2.6)–(2.10) will soon show that only the uniform Airy approximation is easy to apply (i.e. the case $n = 2$). However practical methods for $n = 3$ and $n = 4$ have now been developed, which are reviewed in (Connor 1990). The main problems that must be overcome are (a) accurate and efficient methods are required for the numerical evaluation of the $\{U_k(\zeta_1, \zeta_2, \dots, \zeta_{n-1})\}$ (b) the new parameters $\{\zeta_j\}$ and A in the cuspoid map (2.6) must be determined, and (c) the expansion coefficients $\{q_k\}$ must be found.

2.3   Uniform hyperbolic umbilic approximation

The uniform hyperbolic umbilic approximation can be applied to two-dimensional integrals of the form

$$I_2(\alpha) = \int_{-\infty}^{\infty} \int_{-\infty}^{\infty} g(x,y) \exp[if(\alpha;x,y)/\hbar]\,dx\,dy, \quad \hbar \to 0, \quad (2.11)$$

Typically there are three parameters which control the coalescence of four saddle points. The topology of these saddle points must be such that $f(\alpha;x,y)$ can be mapped onto the hyperbolic umbilic canonical form (Poston & Stewart 1978; Connor 1976)

$$
\begin{aligned}
f(\alpha;x,y) &= u^3 + v^3 + \zeta_1 u + \zeta_2 v + \zeta_3 uv + A & (2.12)\\
&= \Phi(\boldsymbol{\zeta};u,v) + A, & (2.13)
\end{aligned}
$$

say, where $u = u(\alpha;x,y)$, $v = v(\alpha;x,y)$, and the new parameters $\{\zeta_j\}$ and A are just functions of α. As for the cuspoid transformation (2.6), equation (2.12) defines an exact local one-to-one uniformly analytic change of variables provided the saddle points correspond

$$(x_i, y_i) \longleftrightarrow (u_i, v_i), \quad i = 1,2,3,4.$$

Substituting equation (2.13) into the integral (2.11) gives

$$I_2(\alpha) = e^{iA/\hbar} \int_{-\infty}^{\infty} \int_{-\infty}^{\infty} G(u,v) \exp[i\Phi(\boldsymbol{\zeta};u,v)/\hbar]\,du\,dv, \quad (2.14)$$

where

$$G(u,v) = g(x(u,v), y(u,v))J(u,v),$$

and $J(u,v)$ is the Jacobian of the transformation.

The lowest-order terms in the uniform asymptotic approximation are obtained (Connor 1973a, b) by expanding $G(u,v)$ in the form

$$G(u,v) \approx q_{00} + q_{10}u + q_{01}v + q_{11}uv. \quad (2.15)$$

Combining equations (2.14) and (2.15) then allows the hyperbolic umbilic approximation to be written

$$
\begin{aligned}
I_2(\alpha) \quad \sim \quad & \exp(iA/\hbar)\hbar^{2/3}\\
& \times \sum_{k=0}^{1}\sum_{\ell=0}^{1} \hbar^{(k+\ell)/3} q_{k\ell} U_{k\ell}(\hbar^{-2/3}\zeta_1, \hbar^{-2/3}\zeta_2, \hbar^{-1/3}\zeta_3), (2.16)
\end{aligned}
$$

where

$$U_{k\ell}(\zeta_1,\zeta_2,\zeta_3) \equiv \int_{-\infty}^{\infty} \int_{-\infty}^{\infty} u^k v^\ell\, e^{i\Phi}\,du\,dv, \quad (2.17)$$

with $k = 0, 1$ and $\ell = 0, 1$. Note that $U_{k\ell}(\cdot)$ for $k \neq 0$ or $\ell \neq 0$ can be written as a first order partial derivative of $U_{00}(\cdot)$.

The equations presented above were first written down by one of the authors (Connor 1973a, b). Later Ursell (1980) rigorously proved that approximation (2.16) indeed represents the leading terms of a uniform asymptotic expansion.

Before the uniform approximation (2.16) can be applied to practical problems, there are three difficulties that must be overcome, which are similar to those occuring in the uniform cuspoid approximation (2.9). They are (a) numerical methods must be devised for the evaluation of $\{U_{k\ell}(\cdot)\}$, (b) the parameters $\{\zeta_j\}$ and A in the umbilic transformation must be found, and (c) the expansion coefficients $\{q_{k\ell}\}$ must be calculated.

These difficulties have been partly overcome by Uzer & Child (1982) and by Uzer *et al.* (1983). At Manchester, we have improved upon the methods of Uzer and co-workers and have also devised new procedures (Connor & Young, unpublished results) for the evaluation of the uniform umbilic approximation (2.16). Details will be reported elsewhere.

Figures 2.1–2.3 show perspective plots of the hyperbolic umbilic canonical integral, $H(x, y, z)$, defined by

$$H(x, y, z) \equiv \frac{1}{(2\pi)^2} \int_{-\infty}^{\infty} \int_{-\infty}^{\infty} e^{i\Psi} \, du \, dv,$$

where

$$\Psi(x, y, z; u, v) = \tfrac{1}{3}u^3 + \tfrac{1}{3}v^3 + xu + yu + zuv.$$

Note that $H(x, y, z)$ has the properties

$$
\begin{aligned}
H(x, y, z) &= H(y, x, z), \\
H(x, y, z) &= H^*(x, y, -z), \qquad \text{and} \\
H(x, y, 0) &= \mathrm{Ai}(x)\mathrm{Ai}(y).
\end{aligned}
$$

Figure 2.1 plots $|H(x, y, z)|$ for $z = 0$, whilst figures 2.2 and 2.3 show the cases $z = 1$ and $z = 2$ respectively. The structure in figures 2.1–2.3 is best understood in terms of the caustic associated with the hyperbolic umbilic catastrophe (Poston & Stewart 1978) which is drawn in figure 2.4. In particular, the number of real saddle points in different regions of (x, y, z) space, which are marked on figure 2.4, provides a way to comprehend the interference structure in the $|H(x, y, z)|$ plots. This is also the case for the swallowtail canonical integral (2.4), as has been been illustrated earlier (Connor, Curtis & Farrelly 1983, 1984).

2.4 Application of the uniform hyperbolic umbilic approximation

This section illustrates the uniform hyperbolic umbilic approximation by applying it to a test case, namely the integral

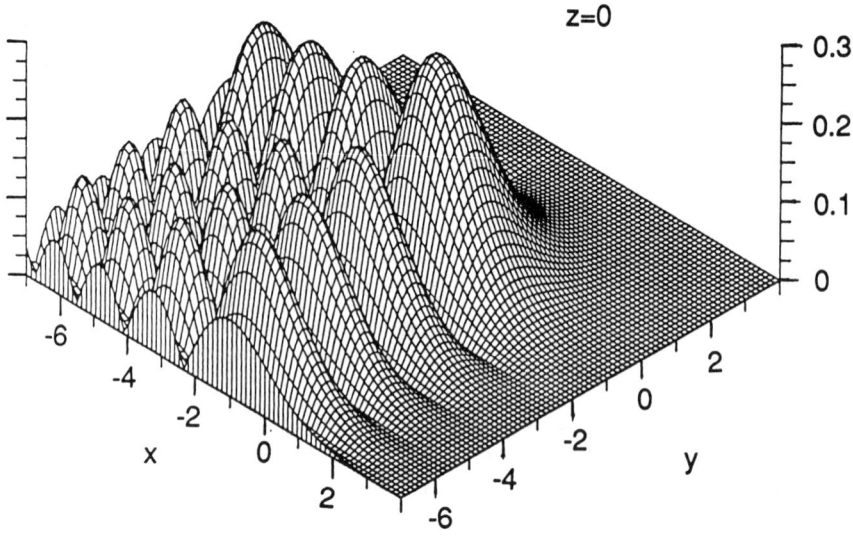

Figure 2.1: Perspective plot of $|H(x, y, z)|$ for $z = 0$.

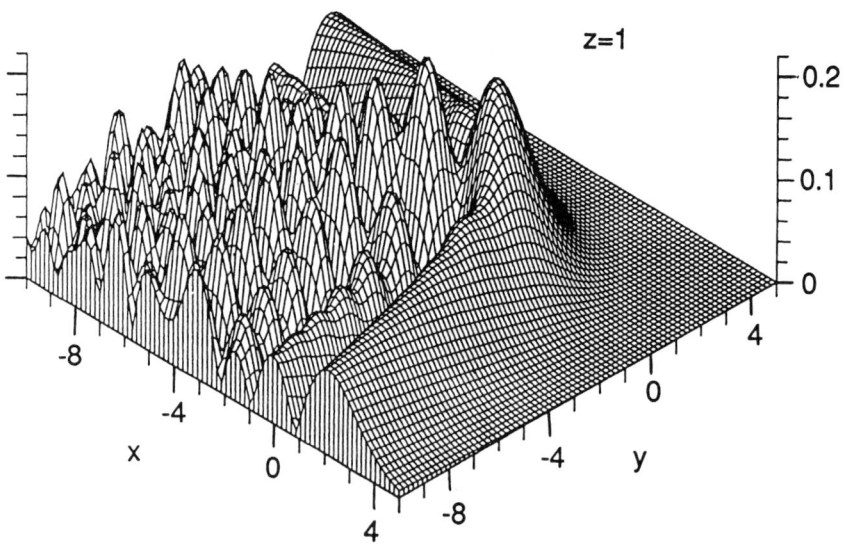

Figure 2.2: Perspective plot of $|H(x, y, z)|$ for $z = 1$.

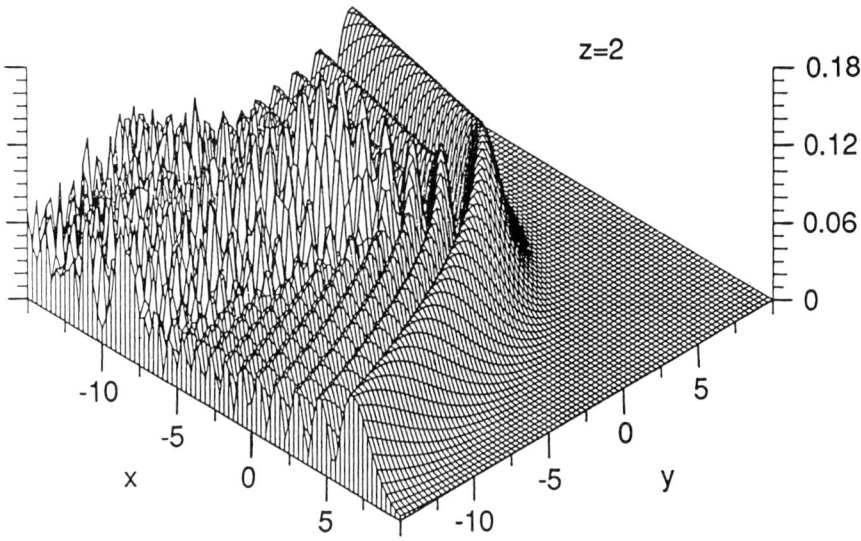

Figure 2.3: Perspective plot of $|H(x, y, z)|$ for $z = 2$.

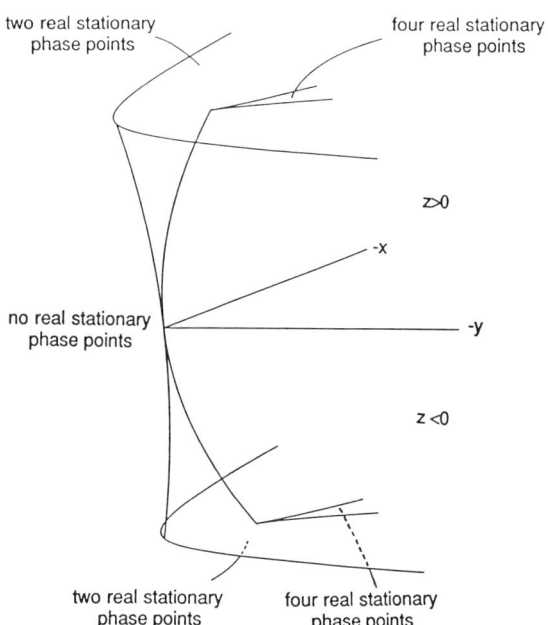

Figure 2.4: Caustic associated with the hyperbolic umbilic catastrophe.

$$T(p, q, a, b, c) = \int_0^{2\pi} \int_0^{2\pi} e^{i\phi(p,q,a,b,c;s,t)} \, ds \, dt, \qquad (2.18)$$

where

$$\phi(p, q, a, b, c; s, t) = ps + qt + a \cos s + b \cos t + c \cos s \cos t.$$

The form of this test integral is suggested by the theory of atom-surface scattering. Note that for $c = 0$, the two-dimensional integral becomes the product of two one-dimensional integrals.

Table 2.1 compares the results of an accurate two-dimensional quadrature (NAG D01DAF) for $T(p, q, a, b, c)$ with the hyperbolic umbilic approximation. For each case $p = 2$, $c = 0.1$ and the remaining parameters have been chosen so there are 0, 2 or 4 real coalescing saddle points (denoted by N). Table 2.1 shows that the absolute error in the uniform approximation is 0.003–0.03 for $\Re T$, corresponding to a relative error of 2% or better. For $\Im T$, the absolute error is 0.003 or better; note that the relative errors can be large in this case since $|\Im T|$ is small in magnitude.

2.5   Franck-Condon integrals

Franck-Condon integrals of the type

$$I_{FC}(\alpha) = \int_{-\infty}^{\infty} \psi_1(x) \psi_2(x) \, dx \qquad (2.19)$$

are of fundamental importance in the description of many spectroscopic and scattering processes involving diatomic molecules. The wavefunctions $\psi_1(x)$ and $\psi_2(x)$ are solutions of the Schrödinger equation

Table 2.1: Results for the test integral $T(p, q, a, b, c)$. In every case $p = 2$ and $c = 0.1$. The number of real coalescing saddle points is denoted by N.

N	a	b	q	2-D quadrature		Uniform approx.	
				$\Re T$	$\Im T$	$\Re T$	$\Im T$
0	1.01	2	4	−0.155	0.0509	−0.158	0.0515
0	2.01	2	4	−0.478	0.0536	−0.484	0.0534
0	3.01	2	4	−0.656	0.00296	−0.662	0.00211
0	1.01	8	6	1.563	0.0563	1.584	0.0577
2	2.01	4	4	−3.940	0.131	−3.973	0.128
2	3.01	4	4	−5.394	0.00716	−5.415	0.00394
2	2.01	6	6	3.446	−0.102	3.471	−0.101
2	3.01	6	6	4.718	−0.00559	4.731	−0.00348
4	3.01	5	4	−7.502	0.00247	−7.519	−0.000399
4	3.01	6	4	−6.855	−0.00596	−6.860	−0.00697
4	3.01	7	6	6.506	−0.00274	6.519	−0.000795
4	3.01	8	6	6.472	0.00326	6.480	0.00408

$$\frac{d^2\psi}{dx^2} + \frac{2m}{\hbar^2}[E - V(x)]\psi = 0, \tag{2.20}$$

where m is the reduced mass, E is the energy, $V(x)$ is the potential energy and $h \equiv 2\pi\hbar$ is the Planck constant. For a bound-continuum transition, $\psi_1(x)$ is the wavefunction for a bound state, and $\psi_2(x)$ is a scattering (or continuum) wavefunction.

As $\hbar \to 0$, $\psi_1(x)$ and $\psi_2(x)$ are typically rapidly oscillating functions of x in classically allowed regions of space. As a result, there are pronounced interference effects in $|I_{FC}|^2$ which is the observable quantity of interest. Now, the integrand of equation (2.19) is not in the form of equation (2.1), to which the uniform cuspoid approximation applies. A way round this difficulty is to use integral representations for $\psi_1(x)$ and $\psi_2(x)$, so that $I_{FC}(\alpha)$ can be written as a three-dimensional integral, followed by a catastrophe theory analysis of the exponent (Connor 1981).

To illustrate the procedure, suppose $V_1(x)$ is a harmonic oscillator potential with eigenvalue E_1. A photon of energy $E_2 - E_1$ then causes a transition to a linear potential $V_2(x)$ with eigenvalue E_2, as illustrated in figure 2.5.

More specifically, if the linear potential is written in the form

$$V_2(x) = ax + b \quad \text{with} \quad a < 0,$$

then the regular solution of the Schrödinger equation (2.20) is

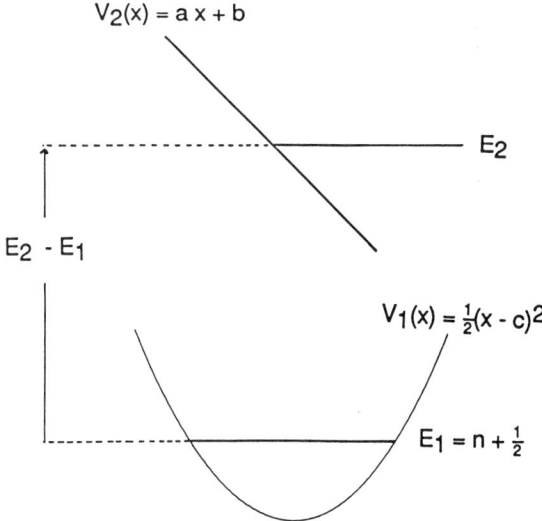

Figure 2.5: The linear-potential harmonic-oscillator model system used for the evaluation of the Franck-Condon integral.

$$\psi_2(x) = \frac{1}{(-a)^{1/6}} \left(\frac{2m}{\hbar^2}\right)^{1/3} \text{Ai}(X_2(x)), \tag{2.21}$$

where

$$X_2(x) = -\left(-a\frac{2m}{\hbar^2}\right)^{1/3}\left(x - \frac{(E-b)}{a}\right).$$

The wavefunction $\psi_2(x)$ has been normalized to an energy delta function. Combining equations (2.2) and (2.21) immediately gives the required integral representation for $\psi_2(x)$.

Next consider the harmonic oscillator potential centered at $x = c$:

$$V_1(x) = \tfrac{1}{2}m\omega^2(x-c)^2, \quad c > 0. \tag{2.22}$$

It is well known that the eigenvalues for the potential (2.22) are $E_n = (n + \tfrac{1}{2})\hbar\omega$, with $n = 0, 1, 2, \ldots$ and the corresponding normalized eigenfunctions are

$$\psi_n(x) = \frac{(m\omega/\hbar)^{1/4}}{(2^n n! \pi^{1/2})^{1/2}} H_n(X_1(x)) \, e^{-\frac{1}{2}X_1^2} \tag{2.23}$$

for $n = 0, 1, 2, \ldots$, where

$$X_1(x) = \left(\frac{m\omega}{\hbar}\right)^{1/2}(x - c),$$

and $H_n(y)$ are Hermite polynomials of degree n. Having defined our notation, it will be convenient to set $\hbar = m = \omega = 1$ from now on.

Unfortunately, equation (2.23) is not in the required integral form. However by relating the Hermite polynomials to Laguerre functions, followed by an application of Cauchy's theorem, Krüger (1981) has derived the following integral representation for the harmonic oscillator wavefunctions

$$\psi_n(x) = \frac{\alpha_n(2n+1)^{1/4}}{2\pi} \int_{-\infty}^{\infty} \frac{\exp[i\mathcal{G}(x,t)]}{(1+t^2)^{3/4}} \, dt, \tag{2.24}$$

for $n = 0, 2, 4, \ldots$, where

$$\mathcal{G}(x,t) = \tfrac{1}{2}\left[(x-c)^2 t - (2n+1)\tan^{-1}t\right]$$

and

$$\alpha_n = \frac{2^{(n+1)/2}\Gamma(1 + \tfrac{1}{2}n)}{[(2n+1)^{1/2}\pi^{1/2}n!]^{1/2}}, \quad n = 0, 1, 2, \ldots,$$

is always close to unity. The integral representation (2.24) is only valid for $n = 0, 2, 4, \ldots$. For $n = 1, 3, 5, \ldots$, a similar integral representation can be derived which has the same exponent, but a different pre-exponential factor (Krüger 1981). In the following we will therefore just consider the case where n is an even integer.

Substituting equations (2.2), (2.21) and (2.24) into $I_{FC}(\alpha)$ gives

$$I_{FC}(\alpha) = A_n \int_{-\infty}^{\infty} \int_{-\infty}^{\infty} \int_{-\infty}^{\infty} \frac{\exp[if(x,s,t)]}{(1+t^2)^{3/4}} \, dx \, ds \, dt, \qquad (2.25)$$

where

$$A_n = \frac{2^{1/3}(2n+1)^{1/4}\alpha_n}{(2\pi)^2(-a)^{1/6}}, \qquad n = 0, 2, 4, \ldots,$$

and

$$f(x,s,t) = \tfrac{1}{3}s^3 - (-2a)^{1/3}[x - (E-b)/a]s + \mathcal{G}(x,t). \qquad (2.26)$$

Note that $f(x,s,t)$ has the symmetry property

$$f(x,s,t) = -f(x,-s,-t). \qquad (2.27)$$

According to equation (2.26), the exponent f depends on four parameters, namely $E - b$, a, n and c. However, we shall keep the quantum number n fixed. It is also convenient for the catastrophe theory analysis to have the most degenerate critical point located at $x = 0$. This has the effect of fixing the value of c at $c = (2n+1)^{1/2}$. Thus f is parametrically dependent on only $E - b$ and a.

2.6   Catastrophe theory analysis of the exponent

In the previous section, we showed that I_{FC} could be written as a three-dimensional integral (2.25), with an exponent defined by equation (2.26). The purpose of this section is to describe how castastrophe theory can be used to analyse the exponent. This allows $f(x,s,t)$ to be locally mapped onto a simpler polynomial form, which is ideally suited to the application of the uniform cuspoid approximation.

The techniques we use are described in Chapter 8 of the monograph by Poston & Stewart (1978). In particular we shall need the following concepts: jet, corank, Splitting Lemma, determinancy, germ, universal unfolding and codimension.

In the rest of this section, we will let $\mathbf{x}$ denote an n dimensional vector, i.e. $\mathbf{x} = (x_1, x_2, \ldots, x_n)$ and will also let $\boldsymbol{\alpha}$ denote a general set of parameters. In addition we assume that the exponent $f(\boldsymbol{\alpha}; \mathbf{x})$ has isolated critical points and is locally equivalent to one of the elementary catastrophes (technically f is 0-modal).

2.6.1   Rules for finding the corank and determinancy of an analytic function

We now present three rules which can be used to find the corank and determinancy (and hence the codimension) of an analytic function $f(\boldsymbol{\alpha}; \mathbf{x})$. We shall need the following standard definition: the k-jet of $f(\mathbf{x})$, denoted $j^k f(\mathbf{x})$, is the Taylor expansion of $f(\mathbf{x})$ about a point $\mathbf{x}_0$ up to and including terms of order k. The point $\mathbf{x}_0$ is usually the origin and will not be indicated explicitly.

STEP 1. Locate the saddle points of $f(\boldsymbol{\alpha}; \mathbf{x})$ or $j^1 f$ by solving $\nabla f = \mathbf{0}$, i.e.,

$$\frac{\partial f}{\partial x_1} = \frac{\partial f}{\partial x_2} = \cdots = \frac{\partial f}{\partial x_n} = 0. \tag{2.28}$$

If $\partial f / \partial x_i \neq 0$ for some i, then f is 1-determinate by the inverse function theorem. However, if equation (2.28) is satisfied for all i, then f is at least 2-determinate. A *degenerate* saddle point occurs when two or more saddle points coalesce, say for $\boldsymbol{\alpha} = \boldsymbol{\alpha}_0$. The number of coalescing saddle points defines the multiplicity of the saddle point.

It is often convenient to perform a linear transformation to remove linear terms from $j^k f$ and to move the degenerate critical point to the origin $\mathbf{x}_0$.

STEP 2. Construct the Hessian matrix $H f(\boldsymbol{\alpha}; \mathbf{x}_0)$ from $f(\boldsymbol{\alpha}; \mathbf{x})$ or $j^2 f$ where

$$H f(\boldsymbol{\alpha}; \mathbf{x}_0) = \left(\frac{\partial^2 f(\boldsymbol{\alpha}; \mathbf{x}_0)}{\partial x_i \partial x_j} \right)$$

with $i = 1, 2, \ldots, n$ and $j = 1, 2, \ldots, n$. Two possibilities now arise depending on whether $\mathbf{x}_0$ is a degenerate saddle point or not:

- If $\det H f \neq 0$, then $\mathbf{x}_0$ is a non-degenerate saddle point. In this case f is 2-determinate and is locally equivalent, for $\mathbf{x}$ close to $\mathbf{x}_0$, to a Morse I-saddle:

$$f(\boldsymbol{\alpha}; \mathbf{x}) = A + e_1 u_1^2 + e_2 u_2^2 + \cdots + e_n u_n^2,$$

where A is a constant, $e_j = \pm 1$ and I, the sum of the negative e_j, determines the type of Morse saddle.

- If $\det H f(\boldsymbol{\alpha}_0; \mathbf{x}_0) = 0$, then $\mathbf{x}_0$ is a degenerate saddle point and f is at least 3-determinate. Next, find the rank, $n-r$, of $H f(\boldsymbol{\alpha}_0; \mathbf{x}_0)$ and hence its corank r. Then, according to the Splitting Lemma, for $\mathbf{x}$ close to $\mathbf{x}_0$, f is equivalent to

$$f(\boldsymbol{\alpha}_0; \mathbf{x}) = A + h(u_1, u_2, \ldots, u_r) + \sum_{j=r+1}^{n} e_j u_j^2, \tag{2.29}$$

where $h(u_1, \ldots, u_r)$ is at least 3-determinate. Equation (2.29) shows that $f(\boldsymbol{\alpha}_0; \mathbf{x})$ depends on r essential variables, with the remaining $n - r$ variables being locally a Morse I-saddle. The corank r is always 1 or 2 for the elementary catastrophes. The correct sign for each e_j, $j = r + 1, r + 2, \ldots, n$ is found by diagonalizing the non-zero part of $H f(\boldsymbol{\alpha}_0; \mathbf{x}_0)$.

STEP 3. Calculate the determinacy of $f(\boldsymbol{\alpha}_0; \mathbf{x})$ at $\mathbf{x}_0$. A function $f(\mathbf{x})$ is k-determinate at $\mathbf{x}_0$ if there is a local equivalence relation of the form

$$f(\mathbf{x}) = j^k f(\mathbf{u}(\mathbf{x})). \tag{2.30}$$

That is, the Taylor expansion of $f(\mathbf{x})$ about $\mathbf{x}_0$ can be safely truncated at order k, before making the change of variables in equation (2.30). The lowest value of k for which f is k-determinate is the *determinacy* of f.

The Mather criterion can be used to find the determinacy of a function in simple cases. This criterion says that if all monomials of degree $k + 1$ in $x_1 \cdots x_n$, i.e. all terms of the type

$$x_1^{\beta_1} x_2^{\beta_2} \cdots x_n^{\beta_n} \quad \text{with} \quad \beta_1 + \beta_2 + \cdots + \beta_n = k + 1,$$

can be written in the form

$$\sum_{i=1}^n \sum_\ell c_{i\ell} j^{k+1} \left(Q_\ell j^{k-1} \left(\frac{\partial (j^k f(\mathbf{x}))}{\partial x_i} \right) \right),$$

where the $c_{i\ell}$ are real numbers and Q_ℓ is any monomial of degree at least 2 in the x_i, then $f(\mathbf{x})$ is k-determinate at $\mathbf{x}_0$ [but may be $(k-1)$-determinate].

If f has been written in the form (2.29), it is only necessary to use the function $h(u_1, \ldots, u_r)$ in the Mather criterion. The smallest k for which the Mather criterion is satisfied may not be the smallest for which f is k-determinate at $\mathbf{x}_0$. However if f has a determinacy of k, then it necessarily satisfies the Mather criterion for $k + 1$. It will be appreciated that checking the Mather criterion can involve considerable algebraic manipulation if k and n are large.

2.6.2 Unfolding the catastrophe

The three rules described in §2.6.1 allow the corank and determinancy of $f(\boldsymbol{\alpha}_0; \mathbf{x})$ at $\mathbf{x}_0$ to be found. Using this information, the appropriate catastrophe germ can be chosen using the list in Table 8.1 of (Poston & Stewart 1978). For umbilic germs, knowing the corank and determinancy will not distinguish between D_4^+ and D_4^- or D_6^+ and D_6^-. However this can usually be done by further examination of the properties of f.

From the calculations in §2.6.1, we now know that for $\mathbf{x}$ close to $\mathbf{x}_0$, $f(\boldsymbol{\alpha}_0; \mathbf{x})$ is locally equivalent to

$$f(\boldsymbol{\alpha}_0; \mathbf{x}) = A + \text{germ}(u_1, \ldots, u_r) + \sum_{j=r+1}^n e_j u_j^2,$$

where $\text{germ}(u_1, \ldots, u_r)$ is one of the standard elementary catastrophe germs, which have a determinacy of ≥ 3. Furthermore, for $(\boldsymbol{\alpha}; \mathbf{x})$ close to $(\boldsymbol{\alpha}_0; \mathbf{x}_0)$, the function $f(\boldsymbol{\alpha}; \mathbf{x})$ is locally equivalent to the universal unfolding of the corresponding catastrophe germ. Formally we can write

$$f(\boldsymbol{\alpha}; \mathbf{x}) = A + \text{cat}(\boldsymbol{\beta}; u_1, \ldots, u_r) + \text{Morse } I\text{-saddle}, \tag{2.31}$$

where $cat(\beta; u_1, \ldots, u_r)$ is the universal unfolding of the catastrophe and the number of new parameters $\beta = \beta(\alpha)$ in the unfolding is the codimension of the singularity.

It may happen that f possesses several nearby degenerate saddle points, which is the case for the exponent (2.26). In this situation, x_0 is chosen as the saddle point of highest multiplicity, since the nearby less degenerate saddle points, as well as the Morse and ordinary (i.e. non-saddle) points are contained in the unfolding of the singularity (2.31). It should also be noted that the mathematically most degenerate singularity that occurs in f may have no physical significance. In this situation, the most degenerate x_0 which still has physical significance is that used in the catastrophe theory analysis.

We have applied the theory described above to the exponent (2.26). Since the details are rather lengthy (albeit straightforward) they will be presented elsewhere. The result of the calculation is that there exists an exact, local, uniform, one-to-one change of variables of the form

$$f(\alpha; x, s, t) = u^5 + \beta_1 u^3 + \beta_3 u + v^2 - w^2. \tag{2.32}$$

The new variables (u, v, w) are functions of the old variables (x, s, t) and the old parameters α. The new parameters β_1 and β_3 are functions of just the old parameters. The terms $\beta_2 u^2$ and A are missing from equation (2.32) because of the symmetry relation (2.27).

2.7   Uniform swallowtail evaluation of Franck-Condon integrals

In §2.6, we showed that the exponent of the three-dimensional Franck-Condon integral (2.25) can be locally mapped onto the form

$$u^5 + \beta_1 u^3 + \beta_3 u + v^2 - w^2,$$

which corresponds to an unfolded swallowtail catastrophe plus a Morse 1-saddle. It is also clear that the Morse term, $v^2 - w^2$, can now be integrated explicitly using the stationary phase method, thereby reducing the three-dimensional integral (2.25) to a one-dimensional integral. This remaining integral can next be evaluated using the uniform swallowtail approximation, as described in §2.2. Thus the overall result is that the original Franck-Condon integral (2.19) for the linear-potential harmonic-oscillator system has a uniform asymptotic expansion in terms of the swallowtail canonical integral (2.4) and its partial derivatives. The techniques reviewed in (Connor 1990) can be used to evaluate numerically this uniform swallowtail approximation.

Figure 2.6 plots $|I_{FC}|^2$ against $E - b$ for the case $n = 6$, $c = (2n + 1)^{1/2}$ and $a = -1.8$. Two curves have been drawn: the solid line, which is the exact result, and the dashed line, which is the uniform swallowtail approximation. The two curves are practically indistinguishable in the figure, showing the high accuracy of the asymptotic treatment. In fact, the relative error is better than 1%, except where $|I_{FC}|^2$ is very small. The absolute error is better than 0.003.

The interference effects in $|I_{FC}|^2$ can be related to the saddle point structure of $f(x, s, t)$. The large peak corresponds to four real saddle points, whereas the oscillations of smaller magnitude typically correspond to $f(x, s, t)$ having two real saddle points. The energy ranges where $|I_{FC}|^2$ is exponentially small correspond to no real saddle points. The physical content of interference effects in Franck-Condon factors has been investigated for many years in the chemical physics literature. However a complete understanding requires uniform asymptotic techniques of the kind described in this chapter.

2.8   Conclusions

The uniform asymptotic approximation of oscillating integrals with coalescing saddle points is a topic of fundamental importance, with many applications to short wavelength scattering phenomen. We have reported two new applications of uniform asymptotic techniques to integrals of the kind that arise in chemical physics. In our first example, we applied the uniform hyperbolic umbilic approximation to a non-separable

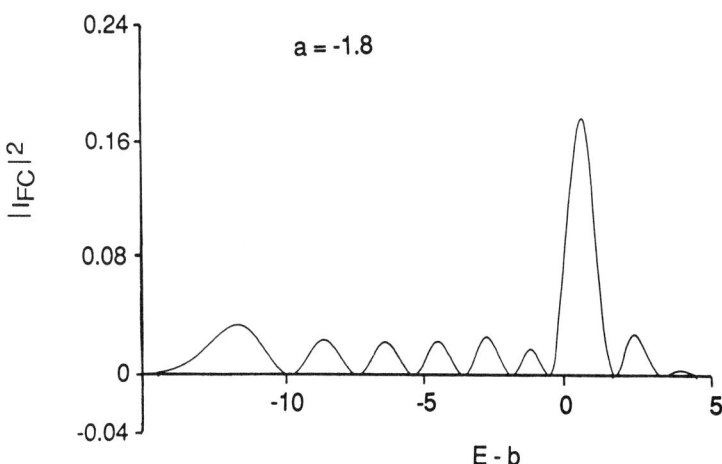

Figure 2.6: Plot of $|I_{FC}|^2$ versus $E - b$ for $n = 6$, $c = (2n + 1)^{1/2}$ and $a = -1.8$. The solid line is the exact result and the dashed line is the uniform swallowtail approximation. These two curves are almost indistinguishable in the figure.

two-dimensional integral suggested by the theory of atom-surface scattering. Our second application concerned Franck-Condon integrals for a linear-potential harmonic-oscillator system. We first wrote the Franck-Condon integral in a three-dimensional form. Next, a catastrophe theory analysis of the exponent was carried out, which showed that it could be mapped locally onto a swallowtail plus a Morse 1-saddle. The resulting uniform swallowtail approximation agreed well with the exact result. This approach allows a uniform asymptotic treatment of an integral to which the usual asymptotic methods do not apply because its integrand is not in standard form.

Acknowledgements

Support of this research by the States of Jersey Education Committee and the UK Science and Engineering Research Council is gratefully acknowledged.

2.9   References

Bleistein, N. 1967 Uniform asymptotic expansions of integrals with many nearby stationary points and algebraic singularities. *J. Math. Mech.* **17**, 533-559.

Bleistein, N. & Handelsman, R.A. 1986 *Asymptotic expansions of integrals.* New York: Dover.

Chester, C., Friedman, B. & Ursell, F. 1957 An extension of the method of steepest descents. *Proc. Camb. Phil. Soc.* **53**, 599-611.

Connor, J.N.L. 1973a Multidimensional canonical integrals for the asymptotic evaluation of the S-matrix in semiclassical collision theory. *Faraday Disc. Chem. Soc.* **55**, 51-53.

Connor, J.N.L. 1973b Evaluation of multidimensional canonical integrals in semiclassical collision theory. *Mol. Phys.* **26**, 1371-1377.

Connor, J.N.L. 1973c Semiclassical theory of molecular collisions: three nearly coincident classical trajectories. *Mol. Phys.* **26**, 1217-1231.

Connor, J.N.L. 1974 Semiclassical theory of molecular collisions: Many nearly coincident classical trajectories. *Mol. Phys.* **27**, 853-866. On p. 861, lines 25 and 26, the words even and odd should be interchanged.

Connor, J.N.L. 1976 Catastrophes and molecular collisions. *Mol. Phys.* **31**, 33-55.

Connor, J.N.L. 1981 Uniform semiclassical evaluation of Franck-Condon factors and inelastic atom-atom scattering amplitudes. *J. Chem. Phys.* **74**, 1047-1052.

Connor, J.N.L. 1990 Practical methods for the uniform asymptotic evaluation of oscillating integrals with several coalescing saddle points. In *Asymptotic and computational analysis* (ed. R. Wong), pp. 137-173. New York: Dekker.

Connor, J.N.L. & Curtis, P.R. 1982 A method for the numerical evaluation of the oscillatory integrals associated with the cuspoid catastrophes: application to Pearcey's integral and its derivatives. *J. Phys. A* **15**, 1179-1190.

Connor, J.N.L. & Curtis, P.R. 1984 Differential equations for the cuspoid canonical integrals. *J. Math. Phys.* **25**, 2895-2902.

Connor, J.N.L. & Farrelly, D. 1981a Molecular collisions and cusp catastrophes: Three methods for the calculation of Pearcey's integral and its derivatives. *Chem. Phys. Lett.* **81**, 306-310.

Connor, J.N.L. & Farrelly, D. 1981b Theory of cusped rainbows in elastic scattering: Uniform semiclassical calculations using Pearcey's integral. *J. Chem. Phys.* **75**, 2831-2846. On p. 2837, 2nd column, line 27, for Fig. 6 read Ref. 6.

Connor, J.N.L., Curtis, P.R. & Farrelly, D. 1983 A differential equation method for the numerical evaluation of the Airy, Pearcey and swallowtail canonical integrals and their derivatives. *Mol. Phys.* **48**, 1305-1330. In Eq. (3.46), for $(4/5)\Gamma(3/5)$ read $(2/5)\Gamma(3/5)$. One line below Eq. (3.50), for $H(x,y,z) = \partial S(x,y,z)/\partial x$ read $H(x,y,z) = \partial S(x,y,z)/\partial z$.

Connor, J.N.L., Curtis, P.R. & Farrelly, D. 1984 The uniform asymptotic swallowtail approximation: practical methods for oscillating integrals with four coalescing saddle points. *J. Phys. A* **17**, 283-310. In Table 4, for $c = -20.957$ read $c = -20.597$.

Fedoriuk, M.V. 1977 *Metod perevala (The saddle point method)*. Moscow: Nauka.

Friedman, B. 1959 Stationary phase with neighboring critical points. *J. Soc. Ind. Appl. Math.* **7**, 280-289.

Krüger, H. 1981 Semiclassical bound-continuum Franck-Condon factors uniformly valid at 4 coinciding critical points: 2 crossings and 2 turning points. *Theoret. Chim. Acta* **59**, 97-116.

Levinson, N. 1961 Transformation of an analytic function of several variables to a canonical form. *Duke Math. J.* **28**, 345-353.

Martin, J. 1974 Integrals with a large parameter and several nearly coincident saddle points: the continuation of uniformly asymptotic expansions. *Proc. Camb. Phil. Soc.* **76**, 211-231.

Olver, F.W.J. 1974 *Asymptotics and special functions*. New York: Academic Press.

Pearcey, T. 1946 The structure of an electromagnetic field in the neighbourhood of a cusp of a caustic. *Phil. Mag.* **37**, 311-317.

Poston, T. & Stewart, I. 1978 *Catastrophe theory and its applications*. London: Pitman.

Ursell, F. 1965 Integrals with a large parameter. The continuation of uniformly asymptotic expansions. *Proc. Camb. Phil. Soc.* **61**, 113-128.

Ursell, F. 1972 Integrals with a large parameter. Several nearly coincident saddle-points. *Proc. Camb. Phil. Soc.* **72**, 49-65.

Ursell, F. 1980 Integrals with a large parameter: a double complex integral with four nearly coincident saddle-points. *Proc. Camb. Phil. Soc.* **87**, 249-273.

Uzer, T. & Child, M.S. 1982 Collision and umbilic catastrophes. Direct determination of the control parameters for use in a uniform S matrix approximation. *Mol. Phys.* **46**, 1371-1388; *erratum ibid.* **50**, 247 (1983).

Uzer, T., Muckerman, J.T. & Child, M.S. 1983 Collisions and umbilic catastrophes. The hyperbolic umbilic canonical diffraction integral. *Mol. Phys.* **50**, 1215-1230.

Wong, R. 1989 *Asymptotic approximation of integrals.* New York: Academic Press.

3

Approximation and asymptotics

Richard E. Meyer
University of Wisconsin

I may be said to have been "present at the creation" because I had the good fortune to share Fritz Ursell's office at a time when he made some of his great discoveries: the reason for the long-wave paradox, the existence of discrete eigenvalues of open water bodies, and the uniform approximation for confluent saddle points. Since I had barely begun to penetrate the surface of my subject at that time, it was an important education to see how much technical mastery, sharp insight and sheer scientific heartache go into major discovery. It gives me much pleasure, therefore, to contribute here a discussion, on which Fritz Ursell has had quite a significant influence, of the role of asymptotics and computation in approximation for scientific issues.

3.1   Introduction

I want to discuss some experiences I have had over the years in obtaining information needed in sciences ranging from aeronautics to oceanography, theoretical chemistry to gas dynamics, meteorology to plasma physics and more. This has involved me in many kinds of approximation issues, and the list of different ideas I have come to exploit, and sometimes to originate, is quite long.

There are three main branches of approximation theory, not disjoint, of course, but distinct enough to have generated separate professions. One is numerical analysis, typically that concerned with difference schemes or finite element methods for the solution of differential equations. Another is approximation by function series related to Hilbert space notions and variational methods, but I shall not talk about this branch now. The third is asymptotics typically exploiting small or large parameters. I have had a lot of experience with that over the last decades, but before that, I was involved heavily and prominently in computing. What has struck me over the years is the degree of analogy between the numerical issues of computation and the analytical issues of asymptotics. This analogy is the main subject I want to talk about. The examples needed are much easier and quicker to display in the more analytical setting of asymptotics, but their relevance to numerical analysis will be apparent without much prompting.

3.2   Objectives and limits

I started out in engineering science and learned a lesson about the objectives of approximation which proved equally valid in all sciences. What they want from applied mathematics is, first of all and most of all, information that illuminates. They want reliable clues to where to go from here. The information needed for that is closer to the purely qualitative than to the accurately quantitative, and here lies the main support for asymptotics. Typical problems of the real world depend on many parameters. Most are plausibly irrelevant and thrown out before the first equation is written down. The most important information by far, that is needed next is how the outcome depends on the parameters left in the problem. This issue is mainly qualitative. The use of computation for it tends to be embarrassing because it is really the last resort when our intelligence must be admitted to have failed. A triumph of applied mathematics arises when it demonstrates that one of the plainly irrelevant parameters is actually decisive; such a discovery changes a whole subject.

Of course, computation is also needed critically. For instance, when a new airplane is to be built, a huge volume of computation is essential. But, the heavy computation is fruitful only after most of the design has been settled. To arrive at the basic design, on the other hand, needs a large volume of approximations to sort out how the outcome depends on the parameters.

I got into computation for the purpose of building experimental equipment intended to serve for a generation, and this is a prime instance where every computational effort to achieve the utmost accuracy is justified ... or so I thought. I learned otherwise when I became responsible for the workshop and was reminded of my apprenticeship as a fitter on the shop floor of a precision-engineering plant. Even the best toolmaker cannot wring five-figure accuracy out of the machining tolerances. That led me to think more seriously about the instrumentation, and the potential for its improvement over the lifetime of even "permanent" laboratory equipment. Spectroscopy is the only method I know that can achieve five-figure accuracy. Everywhere else, the struggle concerns the third significant figure and the fourth is a pipe dream. In the end, I turned to what I should have done first, namely to attempt quantitative estimates of the validity of the differential equations of which I had computed the solutions. I was lucky there because the model I used could match the machining tolerances of three significant figures. The problems of experimental measurement occupied me for a few more years and I published some papers which were too impalatable to be read or quoted. They demonstrated that the experiments mainly sold to Government had great difficulty in nailing down even the second significant figure. . . .

This is how I come to find nearly all computations to more than three significant figures embarrassing. It's not a criticism of computer science because there is a direct analogy in asymptotic expansions. I find them plain embarrassing as a failure of realistic judgment.

In fact, I was led to contemplate an heretical question about asymptotics: are higher approximations than the first justifiable? My experience indicates yes, but rarely. All differential equations are imperfect models and I would be embarrassed to publish a second approximation without convincing justification that the quality of the model validates it.

3.3   Solutions and observables

Applied mathematics has become largely the business of solving differential equations. What bothers me is not only that they are quoted from somewhere and that is judged sufficient reason for solving them. What troubles me even more is the paucity of opportunities for observing the solutions experimentally in any quantitative detail. There are various reasons for it. The solution may describe the detailed behavior of a material that is opaque or inaccessible, as in solid mechanics. Or the medium may be remote, as in ionospheric physics. Again, the medium may be plain contrary, as in geophysics. I have spent years on water waves, which are considered a prime example of classical determinism and observability, because they are easy to see and enjoy. But all serious efforts to measure them accurately in quantitative detail have run into great difficulty.

Quantum mechanics takes this to its logical extreme by declaring that the solutions of Schroedinger's equation are unobservable in principle. The axiom is that only their functionals can be observable. My experience suggests that things are similar in other sciences. Nothing, of course, can be measured except numbers, which must be functionals of the solution, but some functionals are more equal than others. There can be no exhaustive list of them, but typical examples are scattering coefficients of waves or extinction times of populations or most prominently, eigenvalues.

Of course such functionals can be computed from the solutions, but that is not the only way. I shall give examples where it is the worst and dumbest way. The fact is that the solution is determined by the operator of the differential equation and therefore, functionals of the solution must be functionals of the operator itself. Hence, there is no serious reason for a detour via the solutions to approximate what really matters. I have tried to make applied mathematicians more aware of this by publishing a number of examples where the detour via the solution is prohibitive. These examples concern asymptotics, but there is always the analogy: a huge amount of programming, machine time and debugging can be saved by using a small amount of simple analysis to fashion an algorithm for the actually observable quantities as functionals of the operator.

Of course, there are solutions of differential equations, such as Bessel functions, which have achieved a life of their own. It is hard, however, to get away from the fact that it is a somewhat academic life because plainly, no process in the real world satisfies Bessel's equation. Solutions as an end in themselves are pure mathematics; do we really need to know them to eight significant decimals?

3.4   Asymptotic misunderstandings

Over the generations, mathematical asymptotics has come to revolve around the concept of expansion due to Poincaré. He did explain also that they are not complete. This was inconvenient, so the mathematicians formulated a definition of approximation that makes them complete and they built an industry on that definition. Olver (1964) may have been the first to give a simple example of its flaws. He pointed out that

$$\int_0^\pi \frac{\cos nt}{1+t^2}\, dt = I(n^2)$$

has a rigorous asymptotic expansion in powers of n^{-2}. That makes it highly efficient and for $n^2 = 100$, for example, two terms are sufficient to nail the whole expansion down to 4 significant decimals. But, the answer differs by more than 16% from the true value of the integral at $n^2 = 100$. The error exceeds already the second term of the expansion and is accounted for by the term $\frac{1}{2}\pi e^{-n}$ in the integral, which mathematicians consider "totally negligible."

A single example cannot wake up an industry and I was prompted to give a number of further examples indicating that the flaw is not an exception, but is generic in the real world. The simplest such example arises for plane waves in a planely stratified medium. The most common case is that the medium is smooth and the waves are short. Then the solutions of the wave equation have a rigorous asymptotic expansion in the large wave number. But usually, the medium is inaccessible and all one can hope to observe are the total wave transmission and reflection. When people used the solutions to compute reflection, they got paradoxa, no answer, or wrong answers. The reason is that the reflection coefficient has no asymptotic expansion. The mathematicians observed simply that this makes reflection irrelevant by their game rules. However, the wave equation is usually real and then the reflection and transmission coefficients are related exactly by

$$|R|^2 + |T|^2 = 1.$$

This shows that, while the wave solution had been worked out "to all orders", nothing whatever of real relevance in science had been obtained.

Another instructive example concerns waves in a potential shaped like a symmetrical double well, figure 3.1. It had long been conjectured persuasively that the spectrum of such waves is curious in that it consists of pairs of discrete eigenvalues with only a very small split between the members of a pair. Here again, the waves have a rigorous asymptotic expansion in powers of Planck's quantum, and so do the eigenvalues, but the expansion shows all of them to be rigorously degenerate without split. The mathematical theory of this spectrum is largely due to Harrell (1978,1980) and confirms the early conjecture that the split is exponentially small in Planck's quantum.

The significance of this example is illuminated by two remarks. First, spectroscopy observes the split directly, independently of the individual eigenvalues and quite regardless of the fact that it is "zero to all orders". Secondly, the split is very small quantitatively at the lowest eigenvalues, but grows as one proceeds along the spectrum. At the high eigenvalues, the split is about as big as the distance between any two consecutive eigenvalues. How does mathematical asymptotics account for it then? Not at all. The split remains zero quite rigorously according to the definition which Poincaré has been condemned to validate by his name. The expansion is no less rigorous merely because it is wrong by 100%.

Rigour can refer only to deduction from a set of definitions and premises. Unfortunately, it is an axiomatic premise of mathematicians that one cannot be permitted to argue whether a set of consistent definitions might be misplaced.

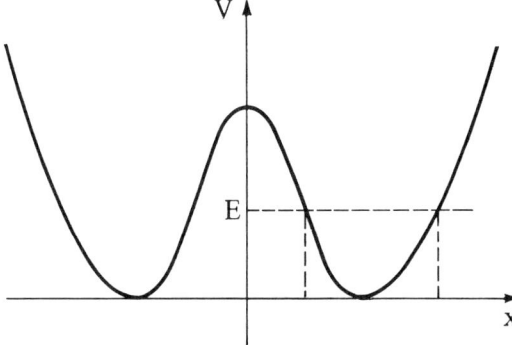

Figure 3.1: Double-well potential

The last decade has seen a growing realization that asymptotic expansions are dubious, but unfortunately, they had already spilled over into other subjects. Numerical analysis talks about third- or fourth-order accuracy with respect to the small parameter of mesh size. This is analogous to the asymptotic notion of accuracy measured in powers of a small parameter. Of course, numerical analysis gives rigorous convergence proofs, which are very similar to those of mathematical asymptotics. Proofs, however, talk about deductions, not premises. The prevalent proofs in numerical analysis proceed from premises that are uncomfortably similar to the premises which have led asymptotics astray.

3.5   Filters

In devising remedies for some such failures of mathematical asymptotics, I have been seeking most of all to give simple remedies. This comes from my view that approximation to exponential precision makes realistic sense only where it estimates quantities which are observable directly. As a corollary, one has a feeling that there ought to be a direct theoretical road to a quantity that is observed directly and that missing such a direct road is somewhat embarrassing. Well, I have been embarrassed, but not always.

The principle of exponential asymptotics is straightforward. Certainly, asymptotic expansions should be avoided because they can't work normally and if they can be made to serve, the method must be excruciatingly clumsy. What is needed is a way to filter what matters out of an approximation mess. The simplest filters, moreover, are likely to be exact filters based on the qualitative structure of the problem.

The very simplest filter I have found works for the reflection of short waves in a stratified medium. It is a natural, brief step to represent the reflection coefficient $|R|$ as the magnitude of a Fourier integral

$$\int_{-\infty}^{\infty} \{[a(x)]^2 - 1\} e^{-2ix/\epsilon} f(x)\, dx, \tag{3.1}$$

with large wave number $2/\epsilon$. It is only a quasi-Fourier integral because $f(x)$ is given, but $a(x)$ is the unknown solution of a nonlinear differential equation. The theory of stationary phase (Jones 1966), however, tells immediately that the integral has no finite critical point when $f(x)$ is quite smooth. (It is physically natural that the points at ∞ should not contribute either to the asymptotic expansion.)

So far, $|R| \sim 0$ again, but the expedient is obvious: shift the path of integration into the lower half of the complex plane (figure 3.2) so that it contains a constant, exponentially small factor $\exp(-2c/\epsilon)$. This needs sharper restrictions on $f(x)$ as $|x| \to \infty$ in the complex plane, but one can be generous first and examine later what was really needed; it turns out to simulate the experimental circumstances to a surprising degree. Naturally, one shifts as far as possible. Then the optimally small exponential factor

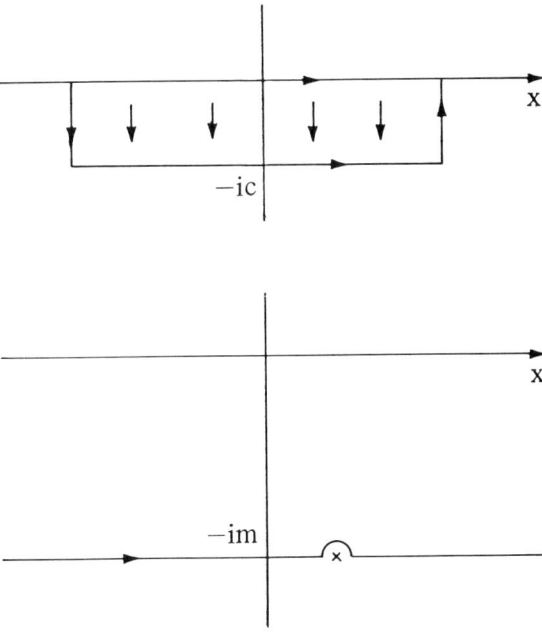

Figure 3.2: Paths for the integral (3.1)

$e^{-m/\epsilon}$ (figure 3.2) has been filtered exactly out of the integral. For the rest, Poincaré asymptotics is adequate and in fact, only the first approximation to the remaining connection problem is usually needed (Meyer 1975).

A similar example from plasma physics concerns the slow modulation of a linear oscillator. The solution is entirely unobservable in the plasma context, but the change of the action over the whole modulation can be observed and has decisive importance in plasma engineering. After long controversies, Littlewood (1963) first showed that the action change has no asymptotic expansion when the modulation is quite smooth. I wanted a concrete estimate. The filter came in two steps. First I transformed to variables akin to action and angle, which made the problem nonlinear, but revealed its structure more clearly. Secondly, I made an existence proof (even though it was mathematically superfluous) to look at the iteration structure. It gave an immediate clue to the definition of a function that both contained the whole asymptotic expansion of the solution and also contributed exactly nothing to the total action change. The remaining Fourier integral could be deformed by steepest descent into a short loop integral.

The simplicity of this exact filter had repercussions. First, I succeeded in treating the slow modulation of an arbitrarily nonlinear Hamiltonian oscillator (Meyer 1976). Then Stengle (1977) was able to dispense with analyticity of the modulation for the linear oscillator and prove a most unexpected transcendental result for the action change. Finally, Leung and Kenneth Meyer (1973) and Stengle (1985) managed to tame the linear Hamiltonian oscillator of any number of degrees of freedom. All those proofs were complicated, but the advances were much too stunning for embarrassment.

The simplicity of those filters stems from the fact that the needed functional has no asymptotic expansion. By contrast, the eigenvalues of short waves in a double well (figure 3.1) have an asymptotic expansion and the effects of tunnelling are hidden behind it, even when they are observable directly. This looks very tough, but leads to a disarmingly simple filter for the double well. Like the experiment, the analysis can be split in two independent parts. One part aims at the degenerate eigenvalue, quite independently of any tunnelling. If the eigenvalue pair is (E_a, E_s), then this part aims to pin down essentially $\frac{1}{2}(E_a + E_s)$ with all sorts of errors, of course, both of the algebraic and transcendental kind. But, those errors have nothing to do with tunnelling. The other part of the analysis studies only the tunnelling. Putting the two parts together gives approximations for E_a and E_s in which the tunnelling effect is swamped by the errors in the first part of the approximation analysis. However, when those poor approximations for E_a and E_s are subtracted, then their joint errors subtract out exactly. The results are separate, rigorous approximations for $\frac{1}{2}(E_a + E_s)$ and $E_a - E_s$, and no more. But, those are just the two quantities which can be observed.

The simplicity of this filter has permitted me to carry exponential asymptotics into the realm of genuine partial differential equations for which the variables cannot be separated (Meyer 1991). That is a much bigger undertaking, but fortunately, almost all the complications are conceptual, rather than analytical, and the computation of the tunnelling split of eigenvalue pairs remains quite simple also in several dimensions.

Exponential asymptotics has also practical applications to continuous spectra, for instance of short waves in a periodic potential (Weinstein & Keller 1985, 1987). I was able to simplify that analysis of Hill's equation drastically by basing it directly on Floquet's original theorem (Magnus & Winkler 1966). It shows the cut-offs of the spectrum to correspond to solutions of a certain periodicity. The filter and proof for the double well then apply almost immediately to the narrow spectral bands. But, when I extended it to several dimensions, I had no Floquet theorem. I did get results, but my method is so complicated and narrow as to embarrass me. Still, it tells us where the basic Floquet theorem goes in partial differential equations.

The filter for the double well rests on the pairing of eigenvalues. Short waves with radiation damping pose the more severe challenge of distinguishing algebraic and transcendental parts of the approximation to just one eigenvalue. Moreover, this problem is not self adjoint and its eigenfunctions are especially contrary. More precisely, the spectrum is real up to a certain point and then it begins to grow complex. The real part of each eigenvalue predicts the frequency and the imaginary part predicts the decay rate of the eigenfunctions. They are observable independently, for instance, spectroscopy measures them independently as line frequency and line width. In some practical respects, the line width is the more important the narrower it is. It will not surprise you to be told that the real part of the eigenvalue has an asymptotic expansion, but the imaginary part is exponential. The filter found by Lozano (Lozano & Meyer 1976) exploits the fact that the very narrow lines occur where the operator is nearly selfadjoint. Exact selfadjointness implies a conjugate complex symmetry of the eigenfunctions which can be traced with somewhat embarrassing labor through the abstract analysis of the contrary eigenfunctions so as to split the characteristic determinant into a real and a complex part. The complex part is exponential and that makes it easy to appeal to the principle of the argument in order to validate the filter.

I was in for an unpleasant surprise when I applied this analysis to theoretical chemistry (Meyer 1986). Connor & Smith (1983) had treated the problem by extremely expensive computation and his formula gave twice my line width. I could not find a misprint in my analysis and Connor's supercomputing program cannot be checked directly, but he is a highly reliable scientist. In the end, I recognized that both of us could be right, for the following reason. The exact formula for an eigenvalue with radiation damping has the structure

$$E = a + t_f + t_d,$$

where a stands for the asymptotic expansion and t_f, for transcendental contributions to the frequency, while t_d denotes the transcendental decay rate. From the point of view of conventional mathematical asymptotics, this formula is rigorously the same as

$$E = a + t_f + \frac{1}{10} t_d$$

because Poincaré asymptotics does not recognize the difference between t_f and $t_f + 9t_d/10$, since both are mere transcendentals hidden behind an expansion. The lesson is that the whole answer for some quantities of practical importance can depend on details of the particular algorithm well below the threshold of what present numerical as well as asymptotic analysis consider of any relevance at all.

I should not close without acknowledging that a proper job in approximation theory requires realistic bounds on the error. In asymptotics, Olver (1974) has pioneered and developed them to a very advanced stage. His bounds, however, are for the uniform approximation of solutions of a class of linear differential equations. Realistic bounds on actually observable functionals are a somewhat different matter and still await investigation.

3.6   Conclusion

Partly due to my efforts, exponential asymptotics has begun to gather the momentum of a bandwagon. This makes me unhappy because it has no relevance of great generality, but brings only a very partial reform.

There are three main points I hope to have communicated in this talk. First, the overwhelming preoccupation of applied mathematics with the solutions of differential equations is not entirely fruitful. Secondly, there are notable analogies between the concerns of numerical and asymptotic analysis and we might benefit from learning more from each other. Most of all, useful approximation cannot be hinged to somebody else's definition, but is quite essentially a matter of informed, realistic judgment.

3.7   References

Connor, J.N.L. & Smith, A.D. 1983 Quantum complex rotation and uniform semiclassical calculations of complex energy eigenvalues. *J. Chem. Phys.* **78**, 6161–6172.

Harrell, E.M. 1978 On the asymptotic rate of eigenvalue degeneracy. *Comm. Math. Phys.* **60**, 73–95.

Harrell, E.M. 1980 Double wells, *Comm. Math. Phys.* **75**, 239–261.

Jones, D.S. 1966 Fourier transforms and the method of stationary phase, *J. Inst. Math. Applics.* **2**, 197–222.

Leung, A.J. & Meyer, K. 1973 Adiabatic invariants for linear Hamiltonian systems, *J. Diff. Eq.* **17**, 32–43.

Littlewood, J.E. 1963 Lorentz's pendulum problem, *Ann. Physics* **21**, 233–242.

Lozano, C. & Meyer, R.E. 1976 Leakage and response of waves trapped by round islands, *Phys. Fluids* **19**, 1075–1088.

Magnus, W. & Winkler, S. 1966 *Hill's equation.* New York: Wiley.

Meyer, R.E. 1975 Gradual reflection of short waves, *SIAM J. Appl. Math.* **29**, 481–492.

Meyer, R.E. 1976 Adiabatic variation, Part V, Nonlinear near-periodic oscillator, *Z. Angew. Math. Phys.* **27**, 181–195.

Meyer, R.E. 1986 Quasiresonance of long life, *J. Math. Phys.* **27**, 238–248.

Meyer, R.E. 1991 On exponential asymptotics for nonseparable wave equations, Parts I–II, to appear in *SIAM J. Appl. Math.*

Olver, F.W.J. 1964 Error bounds for asymptotic expansions. In *Asymptotic solutions of differential equations and their applications.* (ed. C.H. Wilcox), pp. 163–183. New York: Wiley.

Olver, F.W.J. 1974 *Asymptotics and special functions.* New York: Academic Press.

Stengle, G. 1977 Asymptotic estimates for the adiabatic invariance of a simple oscillator, *SIAM J. Math. Anal.* **8**, 640–654.

Stengle, G. 1985 Transcendental estimates for the adiabatic variation of linear Hamiltonian systems, *SIAM J. Math. Anal.* **16**, 932–940.

Weinstein, M.I. & Keller, J.B. 1985 Hill's equation with a large potential, *SIAM J. Appl. Math.* **45**, 200–214.

Weinstein, M.I. & Keller, J.B. 1987 Asymptotic behavior of stability regions for Hill's equation, *SIAM J. Appl. Math.* **47**, 941–958.

4

Converging factors

F.W.J. Olver
University of Maryland

Dedicated, in friendship to Fritz Ursell on the occasion of his retirement from the University of Manchester. His impact on the asymptotic analysis of definite integrals has been, and continues to be, enormous.

An account is provided of the formal researches of J.R. Airey, R.B. Dingle, M.V. Berry and others on converging factors and Stokes' phenomenon, together with recent rigorous analytical investigations of these topics by the present writer. The results aid the theoretical understanding of Stokes' phenomenon and also furnish a basis for the construction of computing algorithms for generating some of the important higher transcendental functions.

4.1 Exponentially accurate asymptotic expansions

Suppose that

$$f(z) \sim f_0 + \frac{f_1}{z} + \frac{f_2}{z^2} + \cdots \tag{4.1}$$

as $z \to \infty$ either along the positive real axis, or more generally from within a certain unbounded region D, say, of the complex plane. Then the error in approximating $f(z)$ by the nth partial sum of the series on the right-hand side of (4.1) is $O(z^{-n})$. Thus we have

$$f(z) = f_0 + \frac{f_1}{z} + \frac{f_2}{z^2} + \cdots + \frac{f_{n-1}}{z^{n-1}} + R_n(z),$$

where

$$|R_n(z)| \le K_n |z|^{-n}, \qquad z \in D,$$

K_n being an assignable constant. This is Poincaré's definition. The integer n is arbitrary, but fixed, and the characteristic feature of the remainder term is that it decays *algebraically* as $z \to \infty$ in D.[1]

---

[1] The infinite series in (4.1) may or may not converge, but this is immaterial as far as Poincaré's definition is concerned.

However, if we permit n to depend on z, then the rate of decay of $R_n(z)$ as $z \to \infty$ can be increased dramatically. For example, consider the integral

$$f(x) = x \int_0^\infty \frac{e^{-xt}}{1+t} dt = xe^x E_1(x), \qquad x > 0, \tag{4.2}$$

$E_1(x)$ being one of two standard notations for the exponential integral. It is well known that

$$f(x) = 1 - \frac{1!}{x} + \frac{2!}{x^2} - \cdots + (-1)^{n-1}\frac{(n-1)!}{x^{n-1}} + R_n(x), \tag{4.3}$$

where

$$0 < (-1)^n R_n(x) < \frac{n!}{x^n}, \qquad x > 0. \tag{4.4}$$

In other words,

$$f(x) \sim \sum_{n=0}^\infty (-1)^n \frac{n!}{x^n}, \qquad x \to \infty, \tag{4.5}$$

and the remainder term is bounded in magnitude by the first neglected term and has the same sign.

For a given value of x, the numerically smallest term in the expansion (4.5) occurs when $n = [x]$, where $[x]$ denotes the integer part of x. (Unless it happens that x is an integer, in which event $n = [x]$ corresponds to one of the two smallest terms.) To minimize the bound for $|R_n(x)|$, we truncate the series at its *optimum stage*, that is, just before the smallest term. Then we have

$$f(x) = \sum_{n=0}^{[x]-1} (-1)^n \frac{n!}{x^n} + R_{[x]}(x), \tag{4.6}$$

where

$$|R_{[x]}(x)| < \frac{[x]!}{x^{[x]}} \sim \sqrt{2\pi x}\, e^{-x} \tag{4.7}$$

as $x \to \infty$, as a consequence of Stirling's formula. Accordingly, $R_n(x)$, with $n = [x]$, decays *exponentially*. We may therefore say that when we neglect $R_{[x]}(x)$ in (4.6) the resulting approximation to $f(x)$ is *exponentially accurate*. (The approximation is not continuous, of course, since the number of terms changes each time x passes through an integer value.)

Can a similar improvement be achieved in the asymptotic expansions of other functions? The answer would appear to be yes, as a rule. However, one of the few general results that we actually have in this direction is the following recent theorem of F. Ursell (1990).[2]

[2]H. Jeffreys (1958) also studied this problem and his result can be related to that of Ursell. However, Jeffreys' analysis is based on complex-variable theory and is heuristic; in contrast Ursell essentially uses real-variable theory and his result is proved rigorously.

Theorem 4.1 *Let $\phi(t)$ be analytic on the disc $|t| < R$, and*

$$\phi(t) = \sum_{n=0}^{\infty} \phi_n t^n, \qquad |t| < R, \tag{4.8}$$

be its Maclaurin expansion. Assume also that $\phi(t)$ is continuous on $[0, \infty)$ and $\phi(t) = O(e^{bt})$ as $t \to \infty$, b being a constant. Then

$$\int_0^{\infty} e^{-xt}\phi(t)dt = \sum_{n \le rx} n! \frac{\phi_n}{x^{n+1}} + O(e^{-rx}) \tag{4.9}$$

as $x \to \infty$, where r is any positive constant less than R.

Here $\sum_{n=0}^{\infty} n!\phi_n x^{-n-1}$ is the asymptotic expansion of the integral on the left-hand side of (4.9) obtained by application of Watson's lemma. The theorem asserts that if we take $[1 + rx]$ terms in the expansion, then the resulting approximation to the integral is exponentially accurate; indeed the error is $O(e^{-rx})$. Ursell's proof is elementary and elegant. It could be extended straightforwardly to sectors in the complex x-plane, and also to cases in which $\phi(t)$ has a singularity at $t = 0$ and the expansion that corresponds to (4.8) involves fractional powers, or logarithms, of t.

4.2   Researches of J.R. Airey

Although there are few general results of the nature of Ursell's that pertain to exponentially-accurate asymptotic expansions, numerous special cases have been investigated, in effect, within the framework of a well-established theory of converging factors. The basic idea in this theory is to truncate an asymptotic expansion either at, or close to, its numerically smallest term, and express the remainder as a multiple of the first neglected term:

$$f(z) = \sum_{s=0}^{n-1} \frac{f_s}{z^s} + C_n(z)\frac{f_n}{z^n},$$

where $n = n(|z|)$ (and we assume that $f_n \ne 0$). Approximations are then sought for the "converging factor" $C_n(z)$ for large $|z|$, or equivalently large n, in cases when the expansion $\sum_{s=0}^{\infty} f_s z^{-s}$ diverges for all finite values of z.

Indeed, this theory goes beyond that proposed in §4.1. The remainder term $C_n(z)f_n z^{-n}$ is again exponentially small for large $|z|$ because the asymptotic expansion is terminated at, or nearly at, the optimum number of terms. But now the coefficient $C_n(z)$ can be calculated accurately from its own approximation, leading to a further improvement in the calculated value of $f(z)$. We may therefore say that the use of converging factors leads to *exponential improvement* in the evaluation of an asymptotic expansion.

The theory can be traced back to T.J. Stieltjes' doctoral dissertation (1886), and has been developed by many researchers, including J.R. Airey (1937), J.C.P. Miller (1952), R.B. Dingle (1958a–f,1973), F.D. Murnaghan & J.W. Wrench Jr. (1963), Murnaghan (1965), Wrench (1970, 1971) and Wrench & V. Alley (1972, 1973). Most of this work has been of a formal nature. Nevertheless, the methods are interesting and the results effective.

Here is Airey's analysis of the exponential integral defined by (4.2), with x replaced by the complex variable z. Since the smallest term in the expansion (4.3) corresponds to $n = [|z|]$, and also to $n = [|z|] - 1$ when $|z|$ is an integer, we set

$$z = \rho e^{i\theta}, \qquad n = \rho - 1 + \alpha, \tag{4.10}$$

with $0 \le \alpha \le 1$, and

$$f(z) = \sum_{s=0}^{n-1} (-1)^s \frac{s!}{z^s} + (-1)^n C_n(z) \frac{n!}{z^n}. \tag{4.11}$$

Then

$$
\begin{aligned}
C_n(z) &= 1 - \frac{n+1}{z} + \frac{(n+1)(n+2)}{z^2} \\
&\quad - \frac{(n+1)(n+2)(n+3)}{z^3} + \cdots,
\end{aligned} \tag{4.12}
$$

that is,

$$
\begin{aligned}
C_n(z) &= 1 - \frac{\rho+\alpha}{\rho}\beta + \frac{(\rho+\alpha)(\rho+\alpha+1)}{\rho^2}\beta^2 \\
&\quad - \frac{(\rho+\alpha)(\rho+\alpha+1)(\rho+\alpha+2)}{\rho^3}\beta^3 + \cdots,
\end{aligned}
$$

where

$$\beta = e^{-i\theta}. \tag{4.13}$$

Airey now rearranges this expansion in descending powers of ρ. The term independent of ρ is given by

$$1 - \beta + \beta^2 - \beta^3 + \cdots = \frac{1}{1+\beta}. \tag{4.14}$$

The coefficient of $1/\rho$ is given by

$$-\alpha\beta + (2\alpha+1)\beta^2 - (3\alpha+3)\beta^3 + \cdots = \frac{\beta^2}{(1+\beta)^3} - \frac{\alpha\beta}{(1+\beta)^2}, \tag{4.15}$$

and so on. Thus

$$
\begin{aligned}
C_n(z) &= \frac{1}{1+\beta} + \left\{ \frac{\beta^2}{(1+\beta)^3} - \frac{\beta}{(1+\beta)^2}\alpha \right\}\frac{1}{\rho} \\
&\quad + \left\{ \frac{\beta^4 - 2\beta^3}{(1+\beta)^5} + \frac{\beta^2 - 2\beta^3}{(1+\beta)^4}\alpha + \frac{\beta^2}{(1+\beta)^3}\alpha^2 \right\}\frac{1}{\rho^2} \\
&\quad + \cdots,
\end{aligned} \tag{4.16}
$$

the coefficient of ρ^{-s} being a polynomial in α of degree s, the coefficients in which are rational functions of β.[3]

That the procedure is formal is obvious from the fact that the series (4.12) diverges, and also from the fact that the series (4.14) and (4.15) converge and are summable (in the ordinary sense) only when $|\beta| < 1$; compare (4.13). That the final result is very effective in practice can be seen from the following calculation extracted from Airey's paper.

Suppose that $z = 5$. Then in (4.10) and (4.11) we have $\rho = 5, \theta = 0$. We may set $n = 4$ or 5. With $n = 4$ we have $\alpha = 0$, and

$$1 - \frac{1!}{5} + \frac{2!}{5^2} - \frac{3!}{5^3} = 0.832, \tag{4.17}$$

exactly, compared with the correct value (Murnaghan & Wrench 1963)

$$f(5) = 5e^5 E_1(5) = 0.85211\,08814\dots\,. \tag{4.18}$$

Accordingly, in its original form the asymptotic expansion yields barely two correct decimal places for this value of z. On evaluating the converging factor from the expansion (4.16), with 23 terms, Airey obtains

$$C_4(5) = 0.52372\,087\dots,$$

Hence, we have

$$1 - \frac{1!}{5} + \frac{2!}{5^2} - \frac{3!}{5^3} + C_4(5)\frac{4!}{5^4} = 0.85211\,08814\dots,$$

in agreement with (4.18) to ten decimal places.

Airey treated several other transcendental functions in a similar vein, including the logarithm of the Gamma function, the confluent hypergeometric function, the incomplete Gamma function, the Laguerre function, and Bessel functions.

The calculations of the coefficients in the asymptotic expansions of the converging factors, typified by (4.16), are apt to be laborious. J.C.P. Miller's contribution (1952) to the theory was to show how to ease these calculations very considerably by ingenious use of differential and difference equations satisfied by the converging factors.

4.3   Researches of R.B. Dingle

In a series of papers, Dingle (1958a–f) formulated a new theory of converging factors, and following in Airey's footsteps he applied this theory to numerous transcendental functions. His expansions are in terms of certain functions related to the generalized exponential integral (or, equivalently, incomplete Gamma function) which he calls *basic converging factors* (or, later, *basic terminants* (Dingle 1973, Chapter 21)). Foremost is the function $\Lambda_p(z)$ defined by

---

[3]Airey calculated the terms up to and including that in $1/\rho^4$. In his notation $\rho = \nu$, $n = r - 1$, $\alpha = -h$. In the case when $\beta = 1$ and $\alpha = 0$ he also calculated the coefficients of $1/\rho^5, 1/\rho^6, \cdots, 1/\rho^{22}$.

$$\Lambda_p(z) = \frac{1}{\Gamma(p+1)} \int_0^\infty \frac{t^p e^{-t}}{1 + (t/z)} \, dt, \tag{4.19}$$

for Re $p > -1$ and $|\text{ph } z| < \pi$. (The relation of $\Lambda_p(z)$ to the generalized exponential integral and the incomplete Gamma function is given below in §4.6.)

The main tool employed by Dingle is Borel's method for summing divergent series. For some of the simpler transcendental functions, including the exponential integral, sine and cosine integrals, error function, Dawson's integral, Fresnel integrals, and the incomplete Gamma function, Borel's method applied to the tail of their asymptotic expansions immediately yields an explicit representation for the remainder term (and hence the converging factor) in terms of $\Lambda_p(z)$ or one of the other basic converging factors. For higher functions, including confluent hypergeometric functions and Bessel functions, the procedure is more complicated. It is first necessary to obtain an integral representation for the remainder term, and then to re-expand this integral as a *series* of $\Lambda_p(z)$ functions.

Here is Dingle's procedure applied to the asymptotic expansion of one of the confluent hypergeometric functions (Dingle 1958d):

$$U(a, c, z) \sim z^{-a} \sum_{s=0}^\infty (-1)^s \frac{(a)_s (1 + a - c)_s}{s! z^s} \tag{4.20}$$

as $z \to \infty$ in the sector $|\text{ph } z| \leq \frac{3}{2}\pi - \delta$ with $\delta > 0$, a and c being real or complex constants. The function $U(a, c, z)$ is defined by

$$U(a, c, z) = \frac{1}{\Gamma(a)} \int_0^\infty t^{a-1} (1 + t)^{c-a-1} e^{-zt} \, dt, \tag{4.21}$$

when Re $a > 0$ and $|\text{ph } z| < \frac{1}{2}\pi$, and by analytic continuation elsewhere. Also, the notation $(a)_s$ stands for, as usual,

$$(a)_s = a(a+1)(a+2) \cdots (a + s - 1);$$

similarly for $(1 + a - c)_s$.

We write

$$U(a, c, z) = z^{-a} \left\{ \sum_{s=0}^{n-1} (-1)^s \frac{(a)_s (1 + a - c)_s}{s! z^s} \right.$$
$$\left. + (-1)^n C_n(z) \frac{(a)_n (1 + a - c)_n}{n! z^n} \right\}, \tag{4.22}$$

n being arbitrary, for the moment. Then from (4.20) we derive

$$C_n(z) \sim \sum_{s=0}^\infty (-1)^s \frac{(n + a)_s (n + 1 + a - c)_s}{(n + 1)_s z^s}$$
$$= \frac{\Gamma(n+1)}{\Gamma(n+a)} \sum_{s=0}^\infty (-1)^s \frac{\Gamma(n + s + a)}{\Gamma(n + s + 1)} \frac{(n + 1 + a - c)_s}{z^s}.$$

Now

$$\frac{\Gamma(n+s+a)}{\Gamma(n+s+1)} = \frac{1}{\Gamma(1-a)} \int_0^\infty \frac{t^{-a}}{(1+t)^{n+s+1}} \, dt, \qquad \text{Re } a < 1.$$

Hence (formally)

$$C_n(z) \sim \frac{\Gamma(n+1)}{\Gamma(1-a)\Gamma(n+a)} \int_0^\infty C_n(t) \, dt, \tag{4.23}$$

where

$$C_n(t) = \sum_{s=0}^\infty \frac{(-1)^s}{z^s} \frac{(n+1+a-c)_s}{t^a(1+t)^{n+s+1}}.$$

However, for large $|z|$

$$\Lambda_{n+a-c}\{(1+t)z\} \sim \sum_{s=0}^\infty (-1)^s \frac{(n+1+a-c)_s}{(1+t)^s z^s}.$$

Hence by comparison with (4.23) we infer that

$$C_n(z) = \frac{n!}{\Gamma(1-a)\Gamma(n+a)} \int_0^\infty \frac{t^{-a}}{(1+t)^{n+1}} \Lambda_{n+a-c}\{(1+t)z\} \, dt$$

(this step constituting the Borel summation procedure). This is the desired integral representation for the converging factor. We now substitute in the integrand by means of the Maclaurin expansion

$$\Lambda_{n+a-c}\{(1+t)z\} = \sum_{s=0}^\infty t^s z^s \Lambda_{n+a-c}^{(s)}(z),$$

and integrate term by term. Here $\Lambda_{n+a-c}^{(s)}(z)$ stands for the reduced derivative

$$\Lambda_{n+a-c}^{(s)}(z) = \frac{1}{s!} \left(\frac{d}{dz}\right)^s \Lambda_{n+a-c}(z).$$

In this way we finally arrive at

$$C_n(z) \sim \sum_{s=0}^\infty \frac{\Gamma(s+1-a)\Gamma(n+a-s)}{\Gamma(1-a)\Gamma(n+a)} z^s \Lambda_{n+a-c}^{(s)}(z), \tag{4.24}$$

or, equivalently,

$$C_n(z) \sim \sum_{s=0}^\infty \frac{(1-a)(2-a)\dots(s-a)z^s \Lambda_{n+a-c}^{(s)}(z)}{(n+a-1)(n+a-2)\dots(n+a-s)}. \tag{4.25}$$

The relation (4.25) is the desired re-expansion of $C_n(z)$. The analysis used in the derivation is purely formal, but Dingle claims that (4.25) is valid when $|\text{ph } z| < \pi$. To apply the result the value of n is taken to be close to $|z|$, so that for large $|z|$ the original expansion (4.20) is truncated at, or close to, its optimum stage. Dingle intended that the basic converging factors $\Lambda_p(z)$ and their reduced derivatives be found from numerical tables, and in Dingle (1958a, 1973), he supplied five-decimal tables of $\Lambda_p(z)$ and the other basic converging factors when z is real and positive and $2p$ is an integer (Murnaghan & Wrench (1963) give further tables of $\Lambda_p(z)$ when p is an integer.) Dingle (1958b, d) also observed that because $\Lambda_p(z)$ and its reduced derivatives are often needed only when p and z are large and fairly close together they themselves can be evaluated from asymptotic expansions, typified by

$$\Lambda_p(p + \xi) \sim \frac{1}{2} \ + \ \frac{2\xi - 1}{8p} + \frac{1 - 2\xi - 4\xi^2}{32p^2}$$
$$+ \ \frac{1 + 6\xi + 8\xi^2 + 8\xi^3}{128p^3} + \cdots$$

as $p \to \infty$, $|\xi|$ being bounded. However, it should be realized that if the basic converging factors and their reduced derivatives are replaced by approximations of this kind in Dingle's expansions for the converging factors, then the results become equivalent to those of Airey.

4.4   Rigorous proofs

What has been described so far, in essence, represented the prevailing state of the art when I wrote my book on asymptotics and special functions in 1965–1974 (Olver 1974). I wished to provide some account of converging factors, and at the same time I wanted the analysis to be rigorous, consistent with the theme of the rest of the book. Eventually I decided to illustrate some of the ideas by means of two examples, the exponential integral $E_1(z)$ and the confluent hypergeometric function $U(a, c, z)$, with $z \to \infty$ in each case. The expansions obtained for the converging factors resemble those of Airey, but the analysis was quite rigorous.

The exponential integral was the easier to treat since there is a simple well-known integral representation for the remainder term in its asymptotic expansion. Following T.J. Stieltjes (1886), J.B. Rosser (1955) and P. Wynn (1963), I obtained results for fixed values of $\theta \equiv \text{ph } z$ in the closed interval $[-\pi, \pi]$ by means of Laplace's method. However, the results are uniform with respect to θ only in compact intervals within the open interval $(-\pi, \pi)$.

The function $U(a, c, z)$ was more difficult, inasmuch as a double integral had to be used to represent the remainder term, and the re-expansion of this double integral required some complicated analysis. The final results are valid for fixed values of θ in the interval $(-\pi, \pi)$, and again uniformity with respect to θ applies only in compact intervals within this range.

4.5   Researches of M.V. Berry

Very recently there has been a surge of renewed interest in the topic of converging factors. D.S. Jones (1990), W.G.C. Boyd (1990) and the present writer (Olver 1990a, b, c) have all been active in deriving new types of expansions using rigorous analyses. In the case of Boyd and myself this work was stimulated by an insightful and important investigation of M.V. Berry (1989).

Berry was interested not in converging factors, *per se*, but in the Stokes phenomenon. For example, in the case of the confluent hypergeometric function $U(a, c, z)$, his approach is to truncate the expansion (4.20) at n terms, and then define a function $S_n(z)$ by the identity

$$
U(a, c, z) \;=\; z^{-a} \sum_{s=0}^{n-1} (-1)^s \frac{(a)_s (1 + a - c)_s}{s! z^s}
$$
$$
+\, S_n(z) 2\pi i \frac{e^{(c - 2a)\pi i}}{\Gamma(a)\Gamma(1 + a - c)} z^{a-c} e^z. \qquad (4.26)
$$

$S_n(z)$ is called the *Stokes multiplier*, and Berry sought an approximation for $S_n(z)$ when z is large and complex, and (4.20) is truncated either at, or close to, the optimum stage. This means n is approximately equal to $|z|$. Berry's goal was to show that if $|z|$ is large and held fixed (so that n, also, can be held fixed), then $S_n(z)$ increases *smoothly* (albeit rapidly) from 0 to 1 as ph z passes continuously from values somewhat below π to values somewhat above π. To put this another way, his objective was to demonstrate that the contribution of the well-known exponentially small term $z^{a-c} e^z$ in the vicinity of ph $z = \pi$ is introduced smoothly, and not as a sudden jump as is commonly supposed. Berry's analysis was purely formal; like that of Dingle it was based in part on Borel summation.

On comparing (4.26) with (4.22), we see that

$$
C_n(z) = (-1)^n \frac{2\pi i e^{(c - 2a)\pi i} n!}{\Gamma(n + a)\Gamma(n + 1 + a - c)} z^{2a - c + n} e^z S_n(z). \qquad (4.27)
$$

Hence any results for the Stokes multiplier $S_n(z)$ can be reinterpreted for the converging factor $C_n(z)$, and *vice versa*. From our standpoint the novel feature of Berry's work is that it leads to uniform asymptotic approximations for $S_n(z)$, and hence $C_n(z)$, in a closed interval of values of ph z that contains π.

After reading Berry's paper I became interested in putting his approach on a rigorous mathematical foundation. This proved to be fairly straightforward in the case of the generalized exponential integral, but more difficult in other cases, including the confluent hypergeometric function. A summary of the results of these investigations forms the subject matter of the next two sections.

4.6 Generalized exponential integral

For real or complex values of the parameter p the generalized exponential integral $E_p(z)$ is defined by

$$E_p(z) = z^{p-1}\Gamma(1-p, z) = e^{-z}\int_0^\infty \frac{e^{-zt}}{(1+t)^p}\, dt \qquad (4.28)$$

when $|\mathrm{ph}\, z| < \frac{1}{2}\pi$, and elsewhere by analytic continuation. The asymptotic expansion

$$E_p(z) \sim \frac{e^{-z}}{z}\sum_{s=0}^\infty (-1)^s \frac{(p)_s}{z^s}, \quad z \to \infty \quad \text{in} \quad |\mathrm{ph}\, z| \le \frac{3}{2}\pi - \delta \qquad (4.29)$$

can be obtained immediately from the integral in (4.28) by applying Watson's lemma (Olver 1974, p. 114). Here δ again denotes an arbitrary positive constant. However, for our purpose it is advantageous to begin instead with the integral representation

$$E_p(z) = \frac{z^{p-1}e^{-z}}{\Gamma(p)}\int_0^\infty \frac{e^{-zt}t^{p-1}}{1+t}\, dt, \qquad (4.30)$$

valid for $\mathrm{Re}\, p > 0$ and $|\mathrm{ph}\, z| < \frac{1}{2}\pi$, which is obtainable from (4.28) by application of the convolution formula for the Laplace transform. (From (4.19) and (4.30) it can be seen that Dingle's basic converging factor is related to the generalized exponential integral via $\Lambda_p(z) = ze^z E_{p+1}(z)$.)

On substituting in (4.30) by means of the identity

$$\frac{1}{1+t} = \sum_{s=0}^{n-1}(-1)^s t^s + (-1)^n \frac{t^n}{1+t}, \quad t \ne -1,$$

and integrating term by term, we arrive at the expansion (4.29) complete with an integral representation of its remainder; thus

$$E_p(z) = \frac{e^{-z}}{z}\sum_{s=0}^{n-1}(-1)^s \frac{(p)_s}{z^s} + (-1)^n \frac{2\pi}{\Gamma(p)}z^{p-1}F_{n+p}(z), \qquad (4.31)$$

where

$$F_{n+p}(z) = \frac{e^{-z}}{2\pi}\int_0^\infty \frac{e^{-zt}t^{n+p-1}}{1+t}\, dt \left(= \frac{\Gamma(n+p)}{2\pi}\frac{E_{n+p}(z)}{z^{n+p-1}}\right), \qquad (4.32)$$

n being arbitrary. Truncation at the optimum number of terms is allowed for by setting

$$z = \rho e^{i\theta}, \quad n = \rho - p + \alpha, \qquad (4.33)$$

with ρ positive and large, and $|\alpha|$ bounded. Since the parameter p may be complex, as long as it is fixed, α may be complex, too. On substituting (4.33) into (4.32) we obtain

$$F_{n+p}(z) = \frac{e^{-z}}{2\pi} \int_0^\infty \frac{e^{-\rho t e^{i\theta}} t^{p+\alpha-1}}{1+t} dt. \tag{4.34}$$

For large ρ, the integrand in (4.34) has a saddle point at $t = e^{-i\theta}$ and a pole at $t = -1$. Thanks to the pioneering work in the 1950's of Ursell and others (see, for example, Chester et al. (1957) and van der Waerden (1951)) we know how to cope with problems of this kind. By introducing a quadratic change of integration variable, we can construct an asymptotic expansion of the right-hand side of (4.34) that is uniformly valid as $\theta \to \pi$, that is, as the pole and saddle-point coalesce. Full details of the analysis are supplied in Olver (1990b). It suffices to observe here that the final expansion for $F_{n+p}(z)$ is uniformly valid when $\theta \in [-\pi + \delta, 3\pi - \delta]$ and $|\alpha|$ is bounded (there is also a similar expansion applicable when $\theta \in [-3\pi + \delta, \pi - \delta]$); furthermore, the leading term is a multiple of the complementary error function. The length, $4\pi - 2\delta$, of the θ-interval of validity is noteworthy, since it exceeds that of the original expansion (4.29). In effect the new expansion embraces two Poincaré-type expansions for $E_p(z)$, and the uniformity with respect to θ fully explains the Stokes phenomenon in the vicinity of $\theta = \pi$ in a quantitative and rigorous manner.

4.7   Confluent hypergeometric function

For other transcendental functions we generally do not have a representation of the remainder term in their asymptotic expansions as simple as that given by (4.32). However (4.32) is a Stieltjes transform, and the first step in other cases is to express the remainder term as a Stieltjes transform, albeit with a nonelementary kernel. In this section we consider the asymptotic expansion of the confluent hypergeometric function $U(a, a - b + 1, z)$ for large $|z|$ and fixed real or complex values of the parameters a and b. The analysis is quite complicated, but the final result is elegant – and simple to describe. Full details of the proof will be found in Olver (1990c).

Define $R_n(a, b, z)$ by the identity

$$U(a, a - b + 1, z) = z^{-a} \sum_{s=0}^{n-1} (-1)^s \frac{(a)_s(b)_s}{s! z^s}$$

$$+ (-1)^n \frac{2\pi z^{b-1} e^z}{\Gamma(a)\Gamma(b)} R_n(a, b, z). \tag{4.35}$$

Suppose that a and b are fixed real or complex numbers and that as $|z| \to \infty$, $|n - |z| + a + b - 1|$ is bounded. Then

$$R_n(a, b, z) = \sum_{s=0}^{m-1} (-1)^s \frac{(1-a)_s(1-b)_s}{s!} \frac{F_{n-s+a+b-1}(z)}{z^s}$$

$$+ (1-a)_m(1-b)_m R_{m,n}(a, b, z), \tag{4.36}$$

where m is an arbitrary fixed nonnegative integer, and

$$R_{m,n}(a, b, z) = O(e^{-z-|z|}z^{-m}), \quad |\mathrm{ph}\, z| \leq \pi, \tag{4.37}$$

$$R_{m,n}(a, b, z) = O(z^{-m}), \quad \pi \leq |\mathrm{ph}\, z| \leq \frac{5}{2}\pi - \delta. \tag{4.38}$$

Once again the total angle, $5\pi - 2\delta$, of the sector of validity is remarkable. In fact the single expansion given by (4.35) to (4.38) embraces the following three expansions of Poincaré type:

$$U(a, a - b + 1, z) \sim z^{-a} \sum_{s=0}^{\infty} (-1)^s \frac{(a)_s(b)_s}{s!z^s}, \tag{4.39}$$

for $|\mathrm{ph}\, z| \leq \frac{3}{2}\pi - \delta$, and

$$U(a, a - b + 1, z) \quad \sim \quad z^{-a} \sum_{s=0}^{\infty} (-1)^s \frac{(a)_s(b)_s}{s!z^s} \tag{4.40}$$

$$\mp \quad 2\pi i \frac{e^{\mp(a+b)\pi i}}{\Gamma(a)\Gamma(b)} \frac{e^z}{z^{1-b}} \sum_{s=0}^{\infty} \frac{(1-a)_s(1-b)_s}{s!z^s}$$

for $|\mathrm{ph}\,(ze^{\mp\frac{3}{2}\pi i})| \leq \pi - \delta$. The essential difference in the structure of (4.35) and (4.36) from these well-known expansions is the presence of the factors $F_{n-s+a+b-1}(z)$ in (4.36). As ph z changes continuously in the various sectors of validity these factors change smoothly to the asymptotic forms required by (4.39) and (4.40).

Exponential improvement (§4.2) occurs throughout the sector $|\mathrm{ph}\, z| \leq \pi$, and falls off gradually as z recedes from this sector. For computational purposes $|\mathrm{ph}\, z| \leq \pi$ is quite sufficient, since connection formulae can be employed elsewhere. Furthermore, sharp bounds on $R_{m,n}(a, b, z)$ have been constructed when $|\mathrm{ph}\, z| \leq \pi$; see Olver (1990c) – and also Boyd (1990) in the case when $U(a, a - b + 1, z)$ reduces to the modified Bessel function $K_\nu(z)$.

Lastly, we observe that there is an obvious resemblance between (4.35) and (4.36) and Dingle's formal expansion given by (4.22) and (4.25). However, there seems to be no simple way to transform one result into the other. Moreover, Dingle claimed only $|\mathrm{ph}\, z| < \pi$ as the region of validity of (4.25), but because of the resemblance of (4.25) to (4.36) it is possible that (4.25) is also valid in the larger sector $|\mathrm{ph}\, z| \leq \frac{5}{2}\pi - \delta$. Perhaps this point is moot: it is already known that the combination of (4.35) and (4.36) *is* valid, in the sense of (4.37) and (4.38), in the larger sector; furthermore, the connection of (4.35) and (4.36) with the Poincaré-type expansions (4.40) is clearer and error bounds are available.

4.8   Conclusions

We have described briefly formal researches on converging factors, especially those of J.R. Airey and R.B. Dingle, together with related recent work of M.V. Berry on the Stokes phenomenon. We have also sketched results due to the present writer and W.G.C. Boyd that place the earlier work, especially that of Berry, on rigorous mathematical foundations. In addition, our new expansions are simpler in form, enjoy more extensive regions of validity, and, in some cases, are accompanied by strict error bounds.

As well as providing further insight into the Stokes phenomenon, the new expansions have a potential in the construction of robust mathematical software for generating many of the higher transcendental functions.

Acknowledgment

This research was supported by the U.S. National Science Foundation under Grant DMS 87-23039.

4.9   References

Airey, J.R. 1937 The "converging factor" in asymptotic series and the calculation of Bessel, Laguerre and other functions. *Phil. Mag.* [7], **24**, 521–552.

Berry, M.V. 1989 Uniform asymptotic smoothing of Stoke's discontinuities. *Proc. Roy. Soc. London, Ser. A* **422**, 7–21.

Boyd, W.G.C. 1990 Stieltjes transforms and the Stokes phenomenon. *Proc. Roy. Soc. London, Ser. A* **429**, 227–246.

Chester, C., Friedman, B. & Ursell, F. 1957 An extension of the method of steepest descents. *Proc. Camb. Phil. Soc.* **53**, 599–611.

Dingle, R.B. 1958a Asymptotic expansions and converging factors I. General theory and basic converging factors. *Proc. Roy. Soc. London, Ser. A* **244**, 456–475.

Dingle, R.B. 1958b Asymptotic expansions and converging factors II. Error, Dawson, Fresnel, exponential, sine and cosine, and similar integrals. *Proc. Roy. Soc. London, Ser. A* **244**, 476–483.

Dingle, R.B. 1958c Asymptotic expansions and converging factors III. Gamma, psi and polygamma functions, and Fermi-Dirac and Bose-Einstein integrals. *Proc. Roy. Soc. London, Ser. A* **244**, 484–490.

Dingle, R.B. 1958d Asymptotic expansions and converging factors IV. Confluent hypergeometric, parabolic cylinder, modified Bessel, and ordinary Bessel functions. *Proc. Roy. Soc. London, Ser. A* **249**, 270–283.

Dingle, R.B. 1958e Asymptotic expansions and converging factors V. Lommel, Struve, modified Struve, Anger and Weber functions, and integrals of ordinary and modified Bessel functions. *Proc. Roy. Soc. London, Ser. A* **249**, 284–292.

Dingle, R.B. 1958f Asymptotic expansions and converging factors VI. Application to physical prediction. *Proc. Roy. Soc. London, Ser. A* **249**, 293–295.

Dingle, R.B. 1973 *Asymptotic expansions: their derivation and interpretation*. London: Academic Press.

Jeffreys, H. 1958 The remainder in Watson's lemma. *Proc. Roy. Soc. London, Ser. A* **248**, 88–92.

Jones, D.S. 1990 Uniform asymptotic remainders. In *Asymptotic and computational analysis* (ed. R. Wong), Lecture Notes in Pure and Applied Mathematics, Vol. 124, pp. 241–264. New York: Marcel Dekker.

Miller, J.C.P. 1952 A method for the determination of converging factors, applied to the asymptotic expansions for the parabolic cylinder functions. *Proc. Camb. Phil. Soc.* **48**, 243–254.

Murnaghan, F.D. 1965 Evaluation of the probability integral to high precision. Report No. 1861, David Taylor Model Basin, U.S. Department of the Navy, Washington, D.C.

Murnaghan, F.D. & Wrench, J.R. 1963 The converging factors for the exponential integral. Report No. 1535, David Taylor Model Basin, U.S. Department of the Navy, Washington, D.C.

Olver, F.W.J. 1974 *Asymptotics and special functions*. New York: Academic Press.

Olver, F.W.J. 1990a On Stokes' phenomenon and converging factors. In *Asymptotic and computational analysis* (ed. R. Wong), Lecture Notes in Pure and Applied Mathematics, Vol. 124, pp. 329–355. New York: Marcel Dekker.

Olver, F.W.J. 1990b Uniform, exponentially-improved, asymptotic expansions for the generalized exponential integral. *SIAM J. Math. Anal.*. [In press.].

Olver, F.W.J. 1990c Uniform, exponentially-improved, asymptotic expansions for the confluent hypergeometric function and other integral transforms. *SIAM J. Math. Anal.*. [In press.].

Rosser, J.B. 1955 Explicit remainder terms for some asymptotic series. *J. Rational Mech. Anal.* **4**, 595–626.

Stieltjes, T.J. 1886 Recherches sur quelques séries semi-convergentes. *Ann. Sci. École. Nor. Sup.* [3], **3**, 201–258. Reprinted in *Complete Works*, Vol. 2, pp. 2–58. Groningen: Noordhoff 1918.

Ursell, F. 1990 Integrals with a large parameter. A strong form of Watson's lemma. In *Elasticity: mathematical methods and applications* (eds. G. Eason & R.W. Ogden), pp. 391–395. Chichester: Ellis Horwood.

van der Waerden, B.L. 1951 On the method of saddle points. *App. Sci. Res., Ser. B* **2**, 33–45.

Wrench, J.W. 1970 The converging factor for the modified Bessel function of the second kind. Report No. 3268, Naval Ship Research and Development Center, U.S. Department of the Navy, Washington, D.C.

Wrench, J.W. 1971 Converging factors for Dawson's integral and the modified Bessel function of the first kind. Report No. 3517, Naval Ship Research and Development Center, U.S. Department of the Navy, Washington, D.C.

Wrench, J.W., & Alley, V. 1972 The converging factor for the sine and cosine integrals. Report No. 3980, Naval Ship Research and Development Center, U.S. Department of the Navy, Bethesda, Maryland.

Wrench, J.W., & Alley, V. 1973 The converging factors for the Fresnel integrals. Report No. 4102, Naval Ship Research and Development Center, U.S. Department of the Navy, Bethesda, Maryland.

Wynn, P. 1963 A numerical study of a result of Stieltjes. *Rev. Française Traitement Informat.* [Chiffres] **6**, 175–196.

5

Matched asymptotic methods in surface wave theory

Frank G. Leppington
Imperial College

It is an honour to be able to contribute to the Fritz Ursell Retirement Meeting. Professor Ursell teaches his research students many things. One point that is made very clear is the importance of rigour. Arguments that are merely plausible can lead to errors, and should therefore be bolstered by rigorous analysis wherever possible. There are certainly dangers in relying on the heuristic rather than the rigorous approach. Nevertheless, there are also distinct advantages in the use of matched expansions which provide asymptotic estimates in a wide class of problems. Such methods are illustrated here with reference to some water wave problems that are associated with earlier work of Fritz Ursell.

5.1   Introduction

The method of matched asymptotic expansions, described succinctly in Van Dyke's book (1964), has had considerable success in dealing with a wide variety of problems, particularly in fluid mechanics. The method does not have full rigour, except in a narrow class of special problems. But the technique does provide a rational procedure in the sense that it can often be used to predict asymptotic approximations to high order, at least in principle.

This talk uses matching ideas in connection with one or two problems in small amplitude water wave theory, in both short and long wave limits. The prototype problems will involve simple scattering geometries, so as to simplify the resulting algebra. The virtue of this approach is that it can be used equally well for more complicated cases and some such results will be quoted in what follows.

Section 5.2 concerns the scattering of short surface waves by a partially immersed cylinder of cross-section S, such that S intersects the free surface vertically, yet with S non-planar at the points of intersection. Details are presented for the simplest case of a semi-circular section S, and some generalisations are given without further details. The related problem of generalisations are given without further details. The related problem of a heaving cylinder of cross-section S is described briefly in §5.3. The case of scattering of short waves by a cylinder, whose cross-section is locally plane and vertical at the intersection points, appears in §5.4.

Finally, an illustration of the long-wave limit is given in §5.5, this concerning the scattering by a circular cylinder of small radius δ, with its central axis immersed to unit depth.

5.2   Transmission past a partially immersed cylinder

The first example (treated in detail by Leppington (1973)) concerns the transmission of short surface waves by a partially immersed cylinder that intersects the free surface vertically; the cross-section S is not plane near the intersection points: the analysis is significantly different for that case, which leads to exponentially small transmitted amplitudes, and this will be treated separately in §5.4.

Cartesian coordinates (x, y) are chosen so that the undisturbed free surface is at $y = 0$, the y-axis points downwards and the scatterer S meets the free surface (FS) at $x = \pm a$. The geometry envisaged is shown in figure 5.1, with intersection points denoted by $P(x = a, y = 0)$ and $Q(x = -a, y = 0)$.

The fluid is forced by an incident wave of potential

$$\Re\left\{\phi_{\text{inc}} e^{-i\omega t}\right\}$$

where $\Re$ denotes the real part of a complex quantity and

$$\phi_{\text{inc}} = \exp\left\{\frac{-i(x - a) - y}{\varepsilon}\right\}.$$

Henceforth the time factor $e^{-i\omega t}$ and the symbol $\Re$ will usually be suppressed.

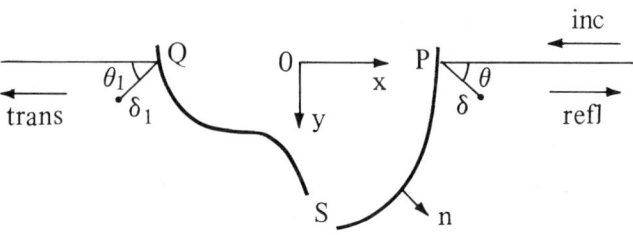

Figure 5.1: Transmission past a partially immersed cylinder.

The velocity potential $\Re\{\phi(x,y)e^{-i\omega t}\}$ is specified by the following equations

$$\nabla^2\phi = 0 \quad \text{in the fluid} \tag{5.1}$$

$$\phi + \varepsilon\phi_y = 0 \quad \text{on } FS \tag{5.2}$$

$$\partial\phi/\partial n = 0 \quad \text{on } S \tag{5.3}$$

$$\phi - \phi_{\text{inc}} \sim \mathcal{R}\, \exp\{i(x+a)/\varepsilon - (y/\varepsilon)\} \quad \text{as } x \to \infty \tag{5.4}$$

$$\phi \sim \mathcal{T}\, \exp\{-i(x-a)/\varepsilon - (y/\varepsilon)\} \quad \text{as } x \to -\infty \tag{5.5}$$

where $\varepsilon = g/w^2$ is a measure of the wavelength, n denotes the outward normal from S and ϕ_y denotes the partial derivative of ϕ with respect to y. The complex reflection and transmission constants $\mathcal{R}$ and $\mathcal{T}$ are unknowns, to be determined. Of particular interest is the asymptotic form of the potential ϕ, and the constants $\mathcal{R}$ and $\mathcal{T}$ in the limit $\varepsilon/a \to +0$.

The solution depends critically on the shape of the scatterer S near the intersection points P and Q. Suppose that near P, S has the form

$$x - a = -f(y) \equiv -\sum_{r=N}^{\infty} \frac{\alpha_r}{r!} y^r \tag{5.6}$$

for $0 \le y < y_0$, where $N \ge 2$. If $N = 2$ then α_N is simply the curvature of S at point P. If $N > 2$, then α_N can be thought of as a generalised curvature there: the larger the value of N, the flatter is the curve S at P. Similarly, near Q we take

$$x + a = f_1(y) = \sum_{r=M}^{\infty} \frac{\beta_r}{r!} y^r, \quad M \ge 2. \tag{5.7}$$

Details are given here for the special case in which S is a semi-circle of radius a, whence

$$M = N = 2 \quad \text{and} \quad \alpha_2 = \beta_2 = 1/a.$$

The asymptotic solution (as $\varepsilon/a \to +0$) is determined by the method of matched expansions. Different forms of approximation are postulated in various overlapping regions, namely: (i) the outer region (at distances large compared with ε from the free surface); (ii) inner regions (within distances $\delta \ll a$ and $\delta_1 \ll a$ from the intersection points P and Q); and (iii) surface wave regions (near the free surface) where surface waves, proportional to $\exp\{(\pm ix - y)/\varepsilon\}$ are significant.

5.2.1 Outer region

Letting $\varepsilon \to +0$ in the specifications (5.1)–(5.5) leads to the hypothesis that ϕ has the form

$$\phi \sim g(\varepsilon)\phi_0$$

where $g(\varepsilon)$ is a scaling factor to be determined and the harmonic function ϕ_0 is subject to the conditions

$$\left.\begin{array}{rll} \partial\phi_0/\partial n = 0 & \text{on} & S \\ \phi_0 = 0 & \text{on} & FS \\ \phi_0 \to 0 & \text{as} & x^2 + y^2 \to \infty. \end{array}\right\} \tag{5.8}$$

These homogeneous conditions admit a non-trivial solution only if ϕ_0 is singular somewhere. It transpires that ϕ_0 has a dipole singularity at the point P.

5.2.2 Right inner region

Near the intersection point $P(x = a, y = 0)$ the dominant length scale is the (small) wavelength parameter ε. Thus, the appropriate scaled variables (X, Y) and Φ are introduced by the formulae

$$x = a + \varepsilon X, \quad y = \varepsilon Y; \quad \phi = \Phi(X, Y).$$

In particular the variables $\delta = ((x - a)^2 + y^2)^{\frac{1}{2}}$ and $R = (X^2 + Y^2)^{\frac{1}{2}}$ are related by the identity

$$\delta = \varepsilon R.$$

In terms of (X, Y) the circle S $(x^2 + y^2 = a^2)$ has the form

$$X = -(\varepsilon/2a)Y^2 - (\varepsilon^3/8a^3)Y^4 \cdots = -F(Y).$$

The boundary condition (5.3) on S is

$$\Phi_X + F'(Y)\Phi_Y = 0 \quad \text{on} \quad X = -F(Y)$$

where suffices denote partial differentiation. It is convenient to transform this condition on to the plane $X = 0$, using Taylor series. For example

$$\Phi_X(X, Y) = \Phi_X(0, Y) + X\Phi_{XX}(0, Y) + \cdots.$$

Thus the right inner potential is the harmonic function Φ subject to the boundary conditions

$$\Phi_X + (\varepsilon/2a)(2Y\Phi_Y - Y^2\Phi_{XX}) + \cdots = 0, \quad X = 0, \ Y > 0, \tag{5.9}$$

$$\Phi + \Phi_Y = 0, \quad Y = 0, \ X > 0, \tag{5.10}$$

together with the radiation requirement

$$\Phi \sim \exp(-iX - Y) + \mathcal{R} \, \exp(iX - Y + 2ia/\varepsilon) \tag{5.11}$$

plus wave-free terms, as $R = (X^2 + Y^2)^{\frac{1}{2}} \to \infty$.

5.2.3 Left inner region

A similar analysis is appropriate for the left inner region, near $Q(x = -a, y = 0)$. Coordinates (X_1, Y_1) are defined by

$$x = -a - \varepsilon X_1, \quad y = \varepsilon Y_1; \quad \phi = \Psi(X_1, Y_1)$$

whence

$$\delta_1 = \varepsilon R_1, \quad \text{with} \quad R_1 = (X_1^2 + Y_1^2)^{\frac{1}{2}}.$$

The harmonic function Ψ is subject to

$$\Psi_{X_1} + (\varepsilon/2a)\left(2Y_1\Psi_{Y_1} - Y_1^2\Psi_{X_1 X_1}\right) + \cdots = 0, \quad X_1 = 0, \ Y_1 > 0,$$

$$\Psi + \Psi_{Y_1} = 0, \quad X_1 > 0, \ Y_1 = 0$$

and

$$\Psi \sim \mathcal{T} \exp(iX_1 - Y_1) \tag{5.12}$$

plus wave-free terms, as $R_1 = (X_1^2 + Y_1^2)^{\frac{1}{2}} \to \infty$.

5.2.4 Matching

The approximations are required to match together smoothly in the various regions of overlap. The wave terms (i.e. those proportional to $\exp((\pm ix - y)/\varepsilon)$ generated in the inner regions are simply continued into the wave region near the free surface. The inner potentials Φ and Ψ will also include wave-free terms (such as $R\sin\theta$) that are required to match with the outer potential ϕ. Roughly,

$$\phi(\delta \to 0) \sim \Phi(R \to \infty) \quad \text{and} \quad \phi(\delta_1 \to 0) \sim \Psi(R_1 \to \infty);$$

that is, the inner limit of the outer potential must be asymptotically equivalent to the outer limit of the inner approximations. The precise conditions are those proposed by Van Dyke (1964); details are given here in Appendix A.

5.2.5 Right inner approximation

It is tentatively assumed that

$$\Phi \sim \Phi_0(X, Y) + \varepsilon\Phi_1(X, Y) + \cdots \tag{5.13}$$

as $\varepsilon/a \to +0$; both Φ_0 and Φ_1 are harmonic. Note that it is *not* assumed that the asymptotic expansion includes only powers of ε, since intermediate terms (for example $\varepsilon^2 \log\varepsilon$), might well occur. A more systematic approach would allow for the possibility of an intermediate term, say $h(\varepsilon)\Phi_1^*$, with $h(\varepsilon)$ such that $\varepsilon \ll h(\varepsilon) \ll 1$, and to be determined by matching. It transpires that there is no such intermediate term in this case (so $\Phi_1^* = 0$) and this step is omitted for the sake of brevity.

Substitution of (5.13) into the specifications (5.9)–(5.11) reveals that

$$\Phi_0 = \exp(-iX - Y) + \exp(iX - Y)$$

and that Φ_1 satisfies the boundary condition

$$\Phi_{1X} = -(1/2a)\left(2Y\Phi_{0Y} - Y^2\Phi_{0XX}\right),$$

i.e.

$$\Phi_{1X} = a^{-1}(2Y - Y^2)e^{-Y} \quad \text{at} \quad X = 0, \ Y > 0$$

and also the free surface condition (5.10). Thus Φ_1 corresponds to a simple wave-maker problem, with solution

$$\Phi_1(X,Y) = \frac{2}{a}\int_0^\infty \mathcal{G}(X,Y;0,Y')\,(2Y' - Y'^2)e^{-Y'}\,dY', \qquad (5.14)$$

where $\mathcal{G}$ is the fundamental source solution, or Green's function. It is given in terms of the function $G(x,y;x',y';K)$, of Appendix B, with $K = 1$. In particular, for large values of R, (5.14) gives

$$\Phi_1 \sim -(i/2a)\exp(iX - Y) + (4/a\pi)R^{-1}\sin\theta.$$

In the notation of Appendix A, the first order wave-free function is

$$\Phi^{(1)} = (4\varepsilon/a\pi)R^{-1}\sin\theta = (4\varepsilon^2/a\pi)\delta^{-1}\sin\theta,$$

thus

$$\Phi^{(1,2)} = (4\varepsilon^2/a\pi)\delta^{-1}\sin\theta. \qquad (5.15)$$

5.2.6   Outer approximation
Expression (5.15) suggests the outer development

$$\phi \sim \varepsilon^2\phi_0 \qquad \text{as} \quad \varepsilon \to 0$$

with

$$\phi_0 \sim (4/a\pi)\delta^{-1}\sin\theta \qquad \text{as} \quad \delta \to 0. \qquad (5.16)$$

Thus the harmonic function ϕ_0 is subject to the boundary condition (5.16), together with (5.8). The solution is readily found (by conformal transformation) as

$$\phi_0 = \Re\left\{\frac{4i}{2\pi a^2}\frac{z + a}{z - a}\right\}$$

where $z = x + iy$. In particular, near $Q(z = -a)$,

$$\phi_0 \sim (\pi a^3)^{-1}\left(\delta_1\sin\theta_1 - (\delta_1^2/2a)\sin 2\theta_1\right)$$

hence

$$\phi^{(2)} \equiv \varepsilon^2\phi_0 = (\pi a^3)^{-1}\left(\varepsilon^3 R_1\sin\theta_1 - (\varepsilon^4/2a)R_1^2\sin 2\theta_1\right). \qquad (5.17)$$

5.2.7 Left outer approximation

Equation (5.17) suggests the left inner form:

$$\Psi \sim \Psi^{(4)} = \varepsilon^3 \Psi_0 + g(\varepsilon)\Psi_1 + \varepsilon^4 \Psi_2 \tag{5.18}$$

where $g(\varepsilon)$ (with $\varepsilon^4 \ll g(\varepsilon) \ll \varepsilon^3$) is to be determined. Matching with $\phi^{(2)}$ requires that

$$\Psi_0 \sim (\pi a^3)^{-1} R_1 \sin \theta_1 = (\pi a^3)^{-1} Y_1$$

and

$$\Psi_2 \sim -(4\pi a^4)^{-1} X_1 Y_1 \quad \text{as} \quad R_1 \to \infty.$$

Also, Ψ_1 can be of order R_1^2 at most as $R_1 \to \infty$ (see Crighton & Leppington (1973)). Thus one finds that

$$\Psi_0 = (\pi a^3)^{-1}(Y_1 - 1)$$

and

$$\Psi_1 = A_1(Y_1 - 1)$$

where A_1 is as yet unknown. The function Ψ_2 is harmonic and subject to the boundary conditions

$$\Psi_{2X_1} = -(\pi a^4)^{-1} Y_1 \quad \text{at} \quad X_1 = 0,$$

$$\Psi_2 + \Psi_{2Y_1} = 0 \quad \text{at} \quad Y_1 = 0;$$

it has the solution

$$\Psi_2 \sim (2i/\pi a^4) \exp(iX_1 - Y_1) \tag{5.19}$$

plus wave-free terms as $R_1 \to \infty$. Finally, the transmission coefficient T is found from (5.19), (5.18) and (5.12) to be

$$T \sim (2i/\pi)(\varepsilon/a)^4 \exp(-2ia/\varepsilon). \tag{5.20}$$

This is the well-known result derived by Ursell (1961). The point of the present method is that it can readily be generalised to deal with more complicated geometries and to obtain higher order approximations if required.

5.2.8 Generalisations

If the outer and inner solutions are expanded to higher order, one finds that

$$T \exp(2ia/\varepsilon) \sim \frac{2i}{\pi}\{(\varepsilon/a)^4 - (4/\pi)(\varepsilon/a)^5 \log(\varepsilon/a) + O(\varepsilon/a)^5\}.$$

If the cross-section S has the more general form described earlier, with its local shape near the intersection points P and Q specified by formulae (5.6), (5.7), then the present method leads to the result

$$\mathcal{T} \exp(2ia/\varepsilon) \sim (8iA/\pi)\varepsilon^{M+N}\alpha_N\beta_M. \tag{5.21}$$

The constant A depends on the global geometry of the scatterer S and is defined implicitly by the following boundary value problem for the outer harmonic potential ϕ_0:

$$\begin{aligned}
\partial\phi_0/\partial n &= 0 \quad \text{on} \quad S \\
\phi_0 &= 0 \quad \text{on} \quad FS \\
\phi_0 &\sim \delta^{-1}\sin\theta \quad \text{as} \quad \delta \to 0 \\
\phi_0 &\sim A\delta_1\sin\theta_1 \quad \text{as} \quad \delta_1 \to 0.
\end{aligned}$$

Some particular examples have been given by Leppington (1973).

Alker (1977) has provided a similar analysis for the case where the scatterer is not vertical at the intersection points. In this case the local inner problems are those of waves at a sloping beach. If S intersects the free surface at external angles $\alpha = (\pi/2\mu)$ at P and $\alpha_1 = (\pi/2\mu_1)$ at Q, Alker showed that

$$\mathcal{T} = O(\varepsilon^{\mu+\mu_1}) \tag{5.22}$$

instead of the result given by (5.20) or (5.21). Note that, in the limit $\alpha \to \pi/2$, $\alpha_1 \to \pi/2$, the estimate (5.22) reduces to $\mathcal{T} = O(\varepsilon^2)$ rather than the $O(\varepsilon^4)$ of equation (5.20). The results are not contradictory, as ε^4 is $O(\varepsilon^2)$, and the explanation is that the coefficient of the term $O(\varepsilon^{\mu+\mu_1})$ vanishes as $\mu \to 1$ or $\mu_1 \to 1$. Alker gives a more accurate result for $\mathcal{T}$, which shows that

$$\mathcal{T} = O\left\{\left(\varepsilon^\mu\sin\mu\pi + \varepsilon^{\mu+1}\right)\left(\varepsilon^{\mu_1}\sin\mu_1\pi + \varepsilon^{\mu_1+1}\right)\right\}$$

in place of (5.22).

5.3   Radiation problem: heaving cylinder

The analysis is somewhat similar for the problem of a vibrating cylinder of cross-section S, and details have been given by Leppington (1973) and Alker (1977). Results are described briefly here for the case in which the cylinder oscillates vertically with speed $\Re\{Ve^{-i\omega t}\}$. Thus the harmonic function ϕ is subject to

$$\phi + \varepsilon\phi_y = 0 \quad \text{on } FS$$

$$\partial\phi/\partial n = V\,\mathbf{j.n} \quad \text{on } S,$$

where $\mathbf{j}$ is the unit vector in the y-direction; also

$$\phi \sim A_\pm\exp\{(\pm ix - y)/\varepsilon\} \quad \text{as} \quad x \to \pm\infty.$$

The outer approximation

$$\phi \sim \phi_0$$

has $\phi_0 = 0$ on the free surface, $\phi_0 \to 0$ as $r \to \infty$ with no singular points (otherwise the solutions could not be matched with inner approximations). The potential ϕ_0 is considered to be known in principle. In particular,

$$\phi_0 \sim AV\delta \sin\theta \quad \text{as} \quad \delta \to 0$$

where A depends on the overall shape of S, so

$$\phi^{(0)} \sim \varepsilon AVR\sin\theta$$

since $\delta = \varepsilon R$. This, taken in conjunction with the inner boundary condition

$$\Phi_X + \varepsilon^{N-1}\frac{\alpha_N}{N!}\left(NY^{N-1}\Phi_Y - Y^N\Phi_{XX}\right) + \cdots$$

$$= \varepsilon^N V\frac{\alpha_N}{(N-1)!}Y^{N-1} + \cdots \quad \text{at} \quad X = 0, \tag{5.23}$$

suggests the right inner development

$$\Phi \sim \varepsilon\Phi_0 + g(\varepsilon)\Phi_1 + \varepsilon^N\Phi_2$$

where $g(\varepsilon)\Phi_1$ might contain one or more intermediate terms with

$$\varepsilon^N \ll g(\varepsilon) \ll \varepsilon.$$

One finds that

$$\Phi_0 = AV(Y-1)$$

and the boundary condition satisfied by Φ_2 is seen from (5.23) to be

$$\frac{\partial\Phi_2}{\partial X} = V(1-A)\frac{\alpha_N}{(N-1)!}Y^{N-1}$$

with

$$\Phi_2 + \Phi_{2Y} = 0 \quad \text{at} \quad Y = 0, \ X > 0.$$

The wave-maker problem for Φ_2 can be solved to give

$$\Phi \sim -2iV(1-A)\alpha_N\exp(iX - Y)$$

as $R \to \infty$ (together with wave-free terms, some of which are large when $R \to \infty$). Thus the wave amplitude at $x = \infty$ is given by

$$A_+ = -2i\varepsilon^N V(1-A)\alpha_N\exp(-ia/\varepsilon).$$

A similar result (of order ε^M) holds for A_-.

For the heaving circular cylinder, for example, $N = 2$, $A = -1$ and $\alpha_N = 1/a$. Other special cases can be determined similarly.

5.4   Scattering from a plane vertical obstacle

The results of §5.2 show that the transmission constant T gets smaller as M and N increase, that is as the scattering surface S gets flatter at the intersection points P and Q.

This section deals with the case (Ayad & Leppington 1977) in which S is plane and vertical near P and Q. It transpires that in such a case T is exponentially small as $\varepsilon \to 0$. Details are given first for the prototype problem wherein S is a rectangular scatterer $(0 < x < 2b, 0 < y < a)$. The geometry envisaged is indicated in figure 5.2, which also shows the local polar coordinate systems (δ, θ) and (δ_1, θ_1) based at the respective corners $(x, y) = (0, a)$ and $(x, y) = (2b, a)$. Extended results are then quoted, without detail, for a more general class of geometries of this type.

With an incident potential $\Re\{\exp((ix - y)/\varepsilon - i\omega t)\}$, it is convenient to express the total potential ϕ_{tot} as

$$\phi_{\text{tot}} = \begin{cases} 2\cos(x/\varepsilon)\exp(-y/\varepsilon) + \phi, & x < 0 \\ \phi, & x > 0 \end{cases} \tag{5.24}$$

where the first term of (5.24) for $x < 0$ corresponds to a totally reflected wave. The harmonic potential ϕ is then subject to the boundary conditions

$$\phi + \varepsilon\phi_y = 0 \quad \text{on} \quad FS$$

$$\partial\phi/\partial n = 0 \quad \text{on} \quad S$$

$$[\phi]_-^+ = 2\exp(-y/\varepsilon), \quad y > a, \tag{5.25}$$

where $[\phi]_-^+$ is the discontinuity $\{\phi(+0, y) - \phi(-0, y)\}$. Finally there are the usual radiation conditions

$$\phi \sim T \exp((ix - y)/\varepsilon) \quad \text{as} \quad x \to \infty, \tag{5.26}$$

$$\phi \sim (\mathcal{R} - 1)\exp((-ix - y)/\varepsilon) \quad \text{as} \quad x \to -\infty.$$

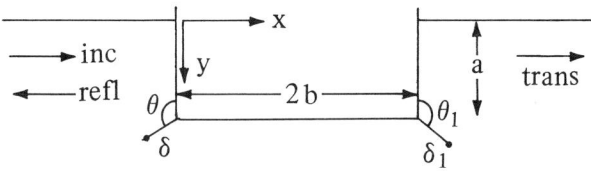

Figure 5.2: Transmission past a rectangular cylinder.

5.4.1   Left inner approximation

In the inner region near the corner at $(x, y) = (0, a)$, one uses coordinates (X, Y) defined by

$$x = \varepsilon X, \quad y = a + \varepsilon Y; \quad \phi(x, y) = \Phi(X, Y)$$

with

$$R = (X^2 + Y^2)^{\frac{1}{2}} = \delta/\varepsilon$$

Equation (5.25) suggests the inner approximation

$$\Phi \sim \exp(-a/\varepsilon)\Phi_0 \tag{5.27}$$

where Φ_0 is harmonic and subject to the conditions

$$[\Phi_0]_-^+ = 2e^{-Y}, \quad X = 0, \quad Y > 0,$$

$$\partial\Phi_0/\partial n = 0 \quad \text{on} \quad S$$

with $\Phi_0 \to 0$ as $R \to \infty$ (otherwise it can not be matched to an outer approximation). The solution for Φ_0 is obtained by using Mellin transforms. In particular it is found that

$$\Phi_0 \sim kR^{-2/3}\cos(2\theta/3) \quad \text{as} \quad R \to \infty \tag{5.28}$$

with

$$k = -(2/\pi)3^{-\frac{1}{2}}\Gamma(2/3)$$

where Γ is the gamma function. On writing the inner approximation (5.27), (5.28) in terms of the outer variable $\delta = \varepsilon R$, one finds

$$\Phi \sim \varepsilon^{2/3}\exp(-a/\varepsilon)k\delta^{-2/3}\cos(2\theta/3),$$

suggesting that the outer approximation (valid not too close to either corner) has the form

$$\phi \sim k\varepsilon^{2/3}\exp(-a/\varepsilon)\phi_0.$$

Thus the harmonic function ϕ_0 satisfies the boundary conditions

$$\partial\phi_0/\partial n = 0 \quad \text{on} \quad S,$$

$$\phi_0 \to 0 \quad \text{as} \quad r \to \infty,$$

$$\phi_0 \sim \delta^{-2/3}\cos(2\theta/3) \quad \text{as} \quad \delta \to 0. \tag{5.29}$$

The solution of this problem, with its prescribed singularity (5.29) at the corner $(x, y) = (0, a)$ can be found using a Schwarz-Christoffel transformation. It is found that near the other corner $(2b, a)$,

$$\phi_0 \sim p + q\delta_1^{2/3}\cos(2\theta_1/3) \quad \text{as} \quad \delta_1 \to 0.$$

The constants p and q are determined in terms of b/a, by an implicit relation involving elliptic integrals, and can be considered to be known in principle.

5.4.2   *Transmission coefficient*

The potential in the region $x > 2b$ is given by the identity

$$\phi(x,y) = 2 \int_a^\infty \mathcal{G}(x,y;2b,y')\, \phi_x(2b,y')\, dy',$$

where $\mathcal{G}$ is the function G of Appendix B, with $K = 1$. It follows from (5.35), (5.29) and (5.26) that

$$\mathcal{T} \sim -2ik\varepsilon^{2/3}\exp(-(a+2ib)/\varepsilon)\int_a^\infty \frac{q\exp(-y'/\varepsilon)}{[3(y'-a)]^{1/3}}\, dy'$$

hence

$$\mathcal{T} \sim (4i/3\pi)(\Gamma(2/3))^2 q\varepsilon^{4/3}\exp(-2(a+ib)/\varepsilon)$$

5.4.3   *Generalisation*

For more general geometries, as shown in figure 5.3, the transmission constant has the form

$$\mathcal{T} \sim F(\lambda,\lambda_1)\,\varepsilon^{\lambda+\lambda_1}\,\exp\{-(a+a_1+2ib)/\varepsilon\},$$

where F depends on the global geometry of S.

5.5   Scattering of long waves by a submerged circular cylinder

Matching methods will be used here for the case of a scatterer whose cross section has small maximum diameter compared with both wavelength and depth.

The prototype problem is taken to be a circular cylinder of (small) radius δ and units of length are chosen so that its centre is at $(0,1)$ with the free surface at $y = 0$ (figure 5.4). The non-dimensional wave number for surface waves is denoted by K and the incident wave has the form $\Re\{\phi_{\text{inc}}(x,y)e^{-i\omega t}\}$, with

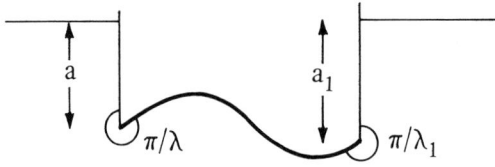

Figure 5.3: Transmission past a plane vertical obstacle.

$$\phi_{\text{inc}} = \exp(K - Ky + iKx)$$

and the factor e^K has been included for later algebraic simplicity. It is also convenient to work with the scattered potential ϕ, defined in terms of the total potential ϕ_{tot} by

$$\phi_{\text{tot}} = \phi_{\text{inc}} + \phi$$

Thus the harmonic function ϕ is subject to the conditions

$$\frac{\partial \phi}{\partial n} = -\frac{\partial \phi_{\text{inc}}}{\partial n} \quad \text{on} \quad S \tag{5.30}$$

$$K\phi + \phi_y = 0 \quad \text{on} \quad FS$$

$$\phi \sim (\mathcal{T} - 1)\exp(K - Ky + iKx) \quad \text{as} \quad x \to \infty$$

$$\phi \sim \mathcal{R}\exp(K - Ky - iKx) \quad \text{as} \quad x \to -\infty.$$

It is required to estimate ϕ, and in particular the constants $\mathcal{T}$ and $\mathcal{R}$, in the limit $\delta \to 0$. (It is known (Dean 1948) that $\mathcal{R} \equiv 0$ for any values of δ and K when the scatterer S is a circular cylinder.)

5.5.1 Outer region

The outer region consists of that part of the fluid domain at distances from S much greater than the radius δ. As $\delta \to 0$ the scattering circle S ($x^2 + (y - 1)^2 = \delta^2$) collapses to a point. The outer potential is that of a line singularity at $(x, y) = (0, 1)$; the nature and strength of this (multipole) singularity has to be determined by matching. It transpires that the leading order approximation ϕ_0 corresponds to a combination of monopole and dipole sources, these being described by the functions G_0, G_1 and G_2 in Appendix B.

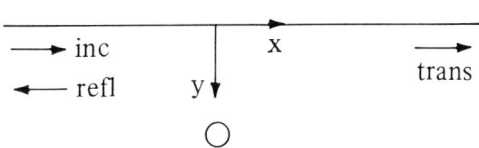

Figure 5.4: Transmission of long waves past a submerged circular cylinder.

5.5.2   Inner approximation

At a point whose distance from S is small compared with the immersion depth and wavelength, the dominant length scale is the radius δ. Thus we introduce rescaled coordinates (X, Y) defined by

$$x = \delta X, \quad y = 1 + \delta Y; \quad \phi = \Phi(X, Y)$$

in terms of which the boundary condition (5.30) at S takes the form

$$\partial \phi / \partial R = -i\delta K e^{i\theta} + \delta^2 K^2 e^{2i\theta} + \cdots$$

when $R = (X^2 + Y^2)^{\frac{1}{2}} = 1$. This suggests the leading order approximation

$$\phi \sim \delta \Phi_1 + \cdots$$

with $\partial \Phi_1 / \partial R = -iK e^{i\theta}$ at $R = 1$. Hence

$$\Phi_1 = iK R^{-1} e^{i\theta} \tag{5.31}$$

unless we include eigen-solutions that are non-vanishing as $R \to \infty$; these are ruled out on the grounds that they cannot be matched with an appropriate outer solution. If (5.31) is rewritten in terms of the outer variable $r = \{x^2 + (y-1)^2\}^{\frac{1}{2}}$, we get

$$\Phi^{(1)} = \delta \Phi_1 = \delta^2 iK r^{-1} e^{i\theta} \tag{5.32}$$

5.5.3   Outer approximation

The form (5.32) suggests

$$\phi \sim \delta^2 \phi_0 \quad \text{as} \quad \delta \to 0$$

with $\phi_0 \sim iK r^{-1} e^{i\theta}$ as $r \to 0$. Hence

$$\phi_0 = -2\pi iK G_1 + 2\pi K G_2 + a_0 G_0 \tag{5.33}$$

where G_0, G_1 and G_2 are described in Appendix B. Leading order matching $\{\Phi^{(1,2)} = \phi^{(2,1)}\}$ does not determine the constant a_0 that multiplies the monopole source potential in formula (5.33), and this has to be evaluated using a higher order matching principle, $\Phi^{(2,2)} = \phi^{(2,2)}$. The form of (5.33) leads to the higher order inner expansion

$$\Phi \sim \delta \Phi_1 + \delta^2 \log \delta \, \Phi_2 + \delta^2 \Phi_3$$

with solutions found to be

$$\Phi_2 = A_2$$

$$\Phi_3 = -\tfrac{1}{2} K^2 R^{-2} e^{2i\theta} + A_3$$

where A_2 and A_3 are constants. The matching principle $\Phi^{(2,2)} = \phi^{(2,2)}$ determines A_2 and A_3 and requires also that $a_0 = 0$. Hence

$$\phi \sim \delta^2 \phi_0 = \delta^2 (2\pi K)(-iG_1 + G_2).$$

In particular the reflection and transmission constants $\mathcal{R}$ and $\mathcal{T}$ are given by

$$\mathcal{T} - 1 \sim 4i\pi\delta^2 K^2 e^{-2K}$$

and

$$\mathcal{R} \sim 0.$$

It has already been pointed out that

$$\mathcal{R} \equiv 0,$$

the result due to Dean (1948) and Ursell (1950a, b) for the totally submerged circular cylinder, for any wavelength and radius. The point of the present method is that it can be generalised to deal with a compressible fluid (when the constant a_0 of formula (5.33) is not zero; see Davis & Leppington (1977)) and for geometries other than the circle (with $\mathcal{R}$ not zero).

5.6   References

Alker, G. 1977 *The radiation and scattering of short surface waves in water*. Ph.D. Thesis, Imperial College, University of London.

Ayad, A.M. & Leppington, F.G. 1977 The diffraction of surface waves by plane vertical obstacles. *J. Fluid. Mech.* **80**, 593–604.

Crighton, D.G. & Leppington, F.G. 1973 Singular perturbation methods in acoustics: diffraction by a plate of finite thickness. *Proc. Roy. Soc.* **A335**, 313–339.

Davis, A.M.J. & Leppington, F.G. 1977 The scattering of electromagnetic surface waves by circular or elliptic cylinders. *Proc. Roy. Soc.* **A353**, 55–75.

Dean, W.R. 1948 On the reflexion of surface waves by a submerged circular cylinder. *Proc. Camb. Phil. Soc.* **44**, 483–491.

Leppington, F.G. 1973 On the radiation and scattering of short surface waves. Part 2. *J. Fluid. Mech.* **59**, 129–146.

Ursell, F. 1950a Surface waves on deep water in the presence of a submerged circular cylinder. I. *Proc. Camb. Phil. Soc.* **46**, 141–152.

Ursell, F. 1950b Surface waves on deep water in the presence of a submerged circular cylinder. II. *Proc. Camb. Phil. Soc.* **46**, 153–158.

Ursell, F. 1961 The transmission of surface waves under surface obstacles. *Proc. Camb. Phil. Soc.* **57**, 638–668.

Van Dyke, M. 1964 *Perturbation methods in fluid mechanics*. New York: Academic Press.

Appendix A. Matching principle

The matching principle that connects the outer expansion $\phi(\delta, \varepsilon)$ and the wave-free part of an inner expansion $\Phi(R, \varepsilon)$, with $\delta = \varepsilon R$, is a modification due to Crighton & Leppington (1973) of the type proposed by Van Dyke (1964). Let $\phi^{(n)}$ denote the outer approximation, up to and including terms of order ε^n, with δ fixed. Rewrite the expression $\phi^{(n)}$ in terms of the inner variable $R = \delta/\varepsilon$ and expand as $\varepsilon \to 0$ with R fixed. Keeping all terms up to those of order ε^m, the result is denoted by $\phi^{(n,m)}$, which is the mth order inner approximation of the nth order outer expansion.

Similarly, the mth order expansion of the wave-free part of the inner potential is denoted by $\Phi^{(m)}$. After rewriting this in terms of the outer variable $\delta = \varepsilon R$, and expanding with δ fixed up to terms of order ε^n, the result is denoted by $\Phi^{(m,n)}$. Van Dyke's matching principle is that

$$\phi^{(n,m)} = \Phi^{(m,n)}$$

for any values of the indices m and n.

Appendix B. Fundamental source solution

The fundamental source solution, or Green's function, is denoted by $G(x, y; x', y'; K)$ and corresponds to the potential at (x, y) induced by a time-harmonic line source at (x', y'). It is specified by the equations

$$\nabla^2 G = \delta(x - x')\delta(y - y'), \qquad y > 0, \ y' > 0, \tag{5.34}$$

where δ is the Dirac delta-function, and

$$G_y + KG = 0 \quad \text{at} \quad y = 0$$

with outgoing waves as $x \to \pm\infty$. The line source singularity at $(x, y) = (x', y')$, implied by the right-hand side of (5.34), leads to the behaviour

$$G \sim (2\pi)^{-1} \log r \quad \text{as} \quad r^2 = (x - x')^2 + (y - y')^2 \to 0.$$

A straightforward Fourier transform analysis leads to the following representation for G:

$$G = (2\pi)^{-1} \log(r/r^*) - i \exp\{iK|x - x'| - K(y + y')\}$$

$$-\frac{1}{\pi} \int_0^\infty \frac{t \cos K(y + y')t - \sin K(y + y')t}{1 + t^2} e^{-K|x-x'|t} \, dt$$

where $r^* = \{(x - x')^2 + (y + y')^2\}^{\frac{1}{2}}$. In particular, one finds that the distant wave form is

$$G \sim -i \exp\{iK|x - x'| - K(y + y')\} \quad \text{as} \quad |x| \to \infty. \tag{5.35}$$

Dipole fields can be generated from G by differentiation with respect to x' or y'. Let $G_0(x, y; K) = G(x, y; 0, 1; K)$ denote the potential due to a line source at $(0, 1)$. Define $G_1 = (\partial/\partial x')G$ and $G_2 = (\partial/\partial y')G$, with (x', y') then set equal to $(0, 1)$. These functions therefore represent vertical and horizontal dipoles at $(0, 1)$. Their near field behaviours are readily found to be

$$G_1 \sim -\frac{\cos\theta}{2\pi r} \quad \text{and} \quad G_2 \sim -\frac{\sin\theta}{2\pi r}, \quad \text{as } r \to 0,$$

where (r, θ) are polar coordinates centred at $(0, 1)$. At great distance from the singularity,

$$G_0 \sim -i\exp(iK|x| - Ky - K)$$

$$G_1 \sim \mp K\exp(iK|x| - Ky - K)$$

$$G_2 \sim iK\exp(iK|x| - Ky - K)$$

as $x \to \pm\infty$.

6

Asymptotics in some biological models

V. Hutson
University of Sheffield

I owe a considerable debt to Professor Fritz Ursell my Ph.D. supervisor, and it is a pleasure to contribute this article to him on the occasion of his retirement. He has inspired the interest of many students in Applied Analysis, an interest which in my case has led to great enjoyment of my research.

6.1   Introduction

6.1.1   General remarks

This contribution is on a completely different subject to the others in this volume, and one which may be unfamiliar to many readers. It does not seem therefore appropriate to present a detailed review of a particular research topic, but rather to outline some of the parts of the relatively new field, mathematical biology, which are being investigated. It should be emphasized that this account is not a review, nor is it intended for the specialist. Rather its aim is to give the general mathematical reader a flavour of some of the fascinating problem areas and of some of the interesting mathematical techniques which are being used to resolve them, and to be very optimistic, to persuade some readers themselves to become involved in research in this area, where there is a host of open problems.

We shall be concerned with problems motivated principally by ecology. In a world suffering severely increasing pressure on the environment from the explosion of the human population, problems of this nature are clearly, it would seem, of rapidly increasing importance, and are likely to become critical in the not too distant future. Nonetheless, as will become clear in the following discussion, the available mathematical techniques are very far from being able to answer most of the questions reasonably posed by biologists. Probably, the relationship of mathematics to biology has not progressed beyond that of mathematics to physics at the beginning of this century. Whilst this may be of concern from a practical point of view it does mean that there is a great opportunity for mathematicians to make progress in this area, and that there is also a great need for major new approaches on some of the extremely difficult problems which arise.

6.1.2 Two mathematical models

It appears that the earliest important work in mathematical ecology was done by Volterra starting about 1925; see (Volterra 1931). Considerable contributions were also made by Lotka (1956), but from a rather different point of view.

Volterra's interest was stimulated by a question asked by the zoologist Umberto d'Ancona, who was later to marry Volterra's daughter. d'Ancona required an explanation for the paradoxical observation that during the first world war, when fishing in the Adriatic was much reduced, the edible fish population *decreased*, while after the war a resumption of fishing caused it to *increase*. Volterra gave the following elegant explanation. He interpreted the total fish population as a mixture of edible fish and predators on them, and with x, y their respective population densities, wrote down the following model:

$$\dot{x} = ax - bxy - kx,$$

$$\dot{y} = -cy + dxy - ky.$$

Here a, b, c, d and k are positive constants, and $\dot{x} \equiv dx/dt$, etc. The constant a represents the per capita difference between birth and death rates of the prey in the absence of the predator, and the term bxy reflects the assumption that the catch rate by the predator is proportional to the product of the prey and predator densities. The last term kx is the depletion rate due to fishing (assumed the same in both prey and predator). The other terms may be interpreted similarly. The pair of equations has an equilibrium at $((c+k)/d, (a-k)/b)$. The population of prey thus *increases* as k the fishing rate increases, an observation that matched the catch statistics. Whether or not this model provides a correct explanation, it initiated an interest in mathematical ecology which particularly in recent years has become very strong. Volterra's book continues to be a fascinating account, containing details of many interesting mathematical techniques and insights based on them.

The ordinary differential equation model which is commonly used is a generalisation of the above idea, and with $u = (u_1, \ldots, u_n)$ the vector of densities in some region, assumed spatially independent, is as follows:

$$\dot{u}_i = u_i f_i(u) \qquad (i = 1, \ldots, n). \tag{6.1}$$

The terms f_i represent the effects of self-interaction and also of interactions from the other species. We sometimes take $g_i(u) = u_i f_i(u)$. Note that the domain is the positive cone $\mathbf{R}_+^n$ and its boundary $\partial\mathbf{R}_+^n$ and interior int $\mathbf{R}_+^n$ are both positively invariant. The simplest assumption is that the f_i are linear, and the following Lotka-Volterra equations are then obtained:

$$\dot{u}_i = u_i\left(a_i - \sum_{j=1}^{n} b_{ij}u_j\right) \quad (i = 1, \ldots, n). \tag{6.2}$$

An extensive discussion of how these and other similar systems of equations arise in ecology, genetics and other contexts may be found in (Hofbauer & Sigmund 1988). The model (6.1) has been much used in ecology. Although there is a host of unanswered questions concerning its behaviour, it is nonetheless clearly a gross oversimplification of reality. For example, the model is autonomous although temporal effects must surely be important. Also, species densities are assumed spatially independent, and age-dependence, delay effects and so on are not included. Here we consider just one generalization and that is to the inclusion of spatial effects. A crude but hopefully illuminating derivation is presented of the simplest model for one species; see (Hoppensteadt 1982).

Take one space dimension and let $u_{m,n}$ be the density at the grid point m at time n. Assume that individuals perform a random walk and move left and right with equal probability $k/2$. Then

$$u_{m,n+1} = \frac{1}{2}ku_{m+1,n} + (1-k)u_{m,n} + \frac{1}{2}ku_{m-1,n}. \tag{6.3}$$

If the grid points are a small distance Δx apart, one can consider the continuous density distribution $u(x,t)$ such that

$$u_{m,n} = u(m\Delta x, \, n\Delta t).$$

Then from (6.3)

$$u(m\Delta x, (n+1)\Delta t) - u(m\Delta x, \, n\Delta t)$$
$$= \frac{1}{2}k[u((m+1)\Delta x, \, n\Delta t) - 2u(m\Delta x, \, n\Delta t)$$
$$+ u((m-1)\Delta x, \, n\Delta t)].$$

By Taylor's theorem

$$u(x, t + \Delta t) = u(x,t) + \frac{\partial u}{\partial t}(x,t)\Delta t + \cdots$$
$$u(x + \Delta x, t) = u(x,t) + \frac{\partial u}{\partial x}(x,t)\Delta x + \frac{1}{2}\frac{\partial^2 u}{\partial x^2}(x,t)\Delta x^2 + \cdots,$$

and it follows on doing the algebra that

$$\frac{\partial u}{\partial t}\Delta t + O(\Delta t^2) = \frac{1}{2}k\frac{\partial^2 u}{\partial x^2}\Delta x^2 + O(\Delta x^3).$$

Letting $\Delta x, \, \Delta t \to 0$ with $\Delta x^2/\Delta t = 1$, one gets

$$\frac{\partial u}{\partial t} = \frac{1}{2}k\frac{\partial^2 u}{\partial x^2}.$$

We shall concentrate here on a generalization of this model to n species, where the basic 'reaction system' is (6.1) and the reaction-diffusion model is obtained as above. With $\Omega \subset \mathbf{R}^m$ an open region with smooth boundary, consider the initial/boundary value problem

$$\frac{\partial u_i}{\partial t} = u_i f_i(u) + \mu_i \Delta u_i, \tag{6.4}$$

$$\frac{\partial u_i}{\partial n} = 0 \quad (x \in \partial\Omega), \tag{6.5}$$

$$u_i(x,0) = u_i(x) \quad (x \in \Omega), \tag{6.6}$$

where Δ is the Laplacian. As usual $\partial/\partial n$ is the normal derivative to the boundary, and (6.5) requires that there is no migration across the boundary, $\partial\Omega$. Other boundary conditions are also of interest, as are other forms of the diffusion, for example degenerate diffusion with terms like $\Delta(u^q)$ with $q > 1$. Often Ω is bounded, but sometimes it is the whole of $\mathbf{R}^m$. For further details concerning the derivation see (Fife 1979) and (Okubo 1980).

6.1.3 *Preliminary observations*

To illustrate some of the difficulties, let us restrict our attention to the system of ordinary differential equations (6.1). This is an n-th order system of nonlinear autonomous equations, and although of course a great deal can be said about the corresponding linear system, the nonlinearities increase the difficulties by several orders of magnitude. In this context one may recall that Hilbert's sixteenth problem, which concerns the number of limit cycles for a pair of such equations with polynomial right-hand sides is largely unsolved. The system (6.1) thus poses a formidable challenge; in this area, according to folklore, $3 = \infty$! Although biological considerations may be expected to restrict the possible range of equations, it does not appear to be known how these restrictions come in.

Granted that this is the case, one might hope perhaps that it is possible to resolve certain restricted questions, for example those concerning the asymptotic (large-time) behaviour. In two dimensions, the 'wildness' of the asymptotic behaviour is limited as is shown by the Poincaré-Bendixson theorem. However, even in three dimensions, it is known that extremely 'bad' asymptotic behaviour is possible. This is exemplified by the famous Lorentz equations, and for Lotka-Volterra models similar conclusions hold (Schaffer *et al.* 1986). One may expect on general grounds, that as the dimension increases, 'bad' behaviour will more and more become the rule rather than the exception.

With the general example of thermodynamics in mind, one might perhaps next ask whether any averaging technique may be profitably employed. However, so far as the author is aware there has been no real progress in this direction. One major snag here is that the system is dissipative, and not Hamiltonian. Nonetheless, this may be the broad

direction in which it is appropriate to look, but some considerable inge-
nuity is probably needed before a breakthrough becomes possible. In this
general context the reader may like to consult (Lasota & Mackey 1985).
The payoff for an averaging technique is suggested by the following simple observation of Volterra, which is unfortunately limited to equations of
Lotka-Volterra type (which otherwise seem not a great deal more tractable
than (6.1)). Suppose a solution $u(t)$ is uniformly bounded above and also
away from the 'faces' $\partial \mathbf{R}_+^n$. In (6.2), divide the i-th equation by u_i, integrate over $(0, T)$ and divide by T to obtain

$$\frac{1}{T} \log \frac{u_i(T)}{u_i(0)} = a_i - \sum_{j=1}^{n} b_{ij}\overline{u}_j, \qquad (6.7)$$

where $\overline{u}_j$ is the time average,

$$\overline{u}_j = \frac{1}{T} \int_0^T u_j(s)ds.$$

Under the above assumption, as $T \to \infty$ the left-hand side of (6.7)
tends to zero, and in the generic case $\det(b) \neq 0$ it follows easily that
$\lim_{T \to \infty} \overline{u}(T) = u^*$, an equilibrium of the system. This yields the rather
striking property that the time average of the orbit approaches an equilibrium. We shall comment on the exploitation of this property in section 6.2.5. It may be added that similar arguments hold for difference
equations (Hofbauer *et al.* 1987) and for differential delay equations (Burton & Hutson 1989).

What then can be done? Certainly, in spite of the rather pessimistic
picture painted above, there has been a great deal of progress on the
mathematical front, and our object here is to give a flavour of some of
this. We start in the next section by asking a question concerning the
gross behaviour of the system, and that is whether long term survival of
the species holds, irrespective of the fine structure of the asymptotics. In
section 6.3, a rather contrasting type of question is asked concerning the
more detailed possibilities for patterns and travelling-waves for reaction-
diffusion systems.

6.2   Permanence

6.2.1   *Introduction*

From the point of view of practical ecology perhaps one of the
most important questions concerns the conditions under which a community of several species will survive over a long period. Analysis of such
questions is all the more important since intuition is a rather poor guide.
To give but one example of this, predation might be thought of as posing
a danger to the survival of some species. However, the opposite can sometimes be true. In a classical experiment a predator was removed from a
marine community of 14 species, and as a result a further 5 species died

out. Thus with a measure of paradox, predation can be a force for the preservation of community structure. The question we shall tackle here is under what conditions on the systems (6.1) and (6.4) (with (6.5) and (6.6)) will coexistence of all species hold.

Clearly we must first ask exactly what is meant by coexistence. The classical approach is to suppose that there is a unique interior (that is in int $\mathbf{R}^n_+$) equilibrium point $\bar{u}$ and to ask that this be stable in some sense. Ideally stable would be interpreted as globally stable (with respect to orbits with initial values in int $\mathbf{R}^n_+$). Volterra invented a clever Liapunov function which works very well for Lotka-Volterra equations with $n = 2$, but gives incomplete answers otherwise. In general it is difficult to discover whether global stability holds even for Lotka-Volterra equations with $n > 2$; an interesting but very hard longstanding problem in this area is the 'Jacobian conjecture', see (Hofbauer & Sigmund 1988, p. 208). A second and weaker notion is that of asymptotic stability. This can be tested by checking the eigenvalues of the Jacobian at $\bar{u}$, and can in principle be done in a wide range of cases. This test has been widely used, and forms the basis of nearly all discussions of coexistence in the literature. Unfortunately, it is only too clear that it is a quite unsatisfactory criterion as the basin of attraction of $\bar{u}$ may be small. In fact it has been argued with increasing force recently that, apart from technical difficulties, global stability is also unsuitable from a biological point of view. This is because many communities have been observed where coexistence appears to hold but the densities are time dependent. One of the best known example of this is the lynx-snowshoe hare interaction. Records of the Hudson Bay Company dating back to 1845 show that very strong cycles in their densities exist with period about ten years. Both species survive in the long term, but the equilibrium is not even asymptotically stable. It is now clear that in a wide variety of fields, including biology and economics, the stability of the system in the sense of its persistence in a broadly unchanged form should not be interpreted as requiring convergence to a time independent state. In two dimensions a limit-cycle, and in higher dimensions even such bad behaviour as a strange attractor seem to occur in some systems which nonetheless persist quite satisfactorily. In an interesting article, Arthur (1990) discusses how in economics this point of view has been surprisingly slow to emerge and has met with considerable resistance.

Such matters have been increasingly under discussion in a biological context recently. About ten years ago a rather appealing definition for coexistence was proposed, and crucially a mathematical technique was devised which allowed the analysis of a reasonably broad range of problems (Schuster *et al.* 1979). The criterion is as follows. The system (6.1) is said to be *permanent* if there are numbers m_1, m_2 with $0 < m_1 \leq m_2$ such that given any $u(0) \in$ int $\mathbf{R}^n_+$ there is T such that

$$m_1 \leq u_i(t) \leq m_2 \quad (1 \leq i \leq n, \, t \geq T).$$

Permanence means that the boundary repels orbits in a uniform sense; however the detailed asymptotic behaviour of the orbits is not relevant, and could be quite 'wild'. Furthermore the criterion is global in that the initial values of the orbit do not matter. There is a strong case for arguing that this criterion reflects the intuitive idea of coexistence much better than the classical ones. The aim of this section is to outline a technique for tackling permanence which has proved relatively successful both with the system of ordinary differential equations (6.1) and the reaction-diffusion system (6.4). In developing the technique we outline some of the key features of dynamical systems theory which enable a unified treatment of a wide range of models including (6.1) and (6.4) to be given. These features are basic in many current lines of attack on nonlinear problems. For a further discussion of permanence, see (Hofbauer & Sigmund 1988) who deal with the ordinary differential equation case, the review by Hutson & Schmitt (1990) which considers this case and also covers in some detail the technicalities required to tackle reaction-diffusion systems, and an outline article by Waltman (1990).

6.2.2   Semiflows

The basic notation and terminology are outlined below; for further details one may consult (Bhatia & Szegö 1970), (Saperstone 1981) and (Hale 1988). Permanence is an asymptotic property, and in essentially infinite dimensional semiflows such as those generated by reaction-diffusion systems, unless some limitation is imposed on the asymptotic behaviour of orbits, it is extremely hard to make much progress with questions in this area. Fortunately, as is well known, for such systems the solution is smoother than the initial data, and this smoothing action is enough to allow the theory to operate.

Let (Y, d) be a metric space. Points in Y will be denoted by $u, v, \ldots$ and subsets of Y by $U, V, \ldots$. There are two notions of distance of sets needed:

$$\overline{d}(U, V) = \sup_{u \in U} d(u, V),$$

$$\underline{d}(U, V) = \inf_{u \in U} d(u, V).$$

Definition 6.1 *The triple* $(Y, \pi, \mathbf{R}_+)$ *is called a semiflow (or semidynamical system) if for all* $u \in Y$, $\pi : Y \times \mathbf{R}_+ \to Y$ *satisfies*

  (i) $\pi(u, 0) = u$,
  (ii) $\pi(\pi(u, t), s) = \pi(u, t + s) \quad (s, t \in \mathbf{R}_+)$,
  (iii) π *is continuous.*

If $u(t)$ is the solution of a system of autonomous ordinary differential equations with $u(0) = u$, one would take $\pi(u, t) = u(t)$. Property (ii), called the semigroup property, simply says that going forward by time t followed by s is the same as going forward a time $(t + s)$. The third property is continuity with respect to the initial conditions and with respect to time. The term 'semiflow' suggests the nice picture of a solution being generated by a fluid flowing forwards in time.

To simplify the notation we often write $\pi(u, t) = u.t$. For $U \subset Y$ and $I \subset \mathbf{R}_+$,

$$U.I = \bigcup_{\substack{t \in I \\ u \in U}} u.t.$$

u is said to be an *equilibrium point* if $u.t = u \ (t \in \mathbf{R}_+)$. The set

$$\gamma^+(u) = \{v : v = u.t \text{ for some } t \in \mathbf{R}_+\}$$

is called the semi-orbit through u, and $\gamma^+(U)$ is defined by taking unions. It is not always the case, for example for reaction diffusion systems, that a solution backwards in time exists. We shall sometimes be interested in situations where, at least for some u, there is a continuation back to $-\infty$. Here we impose the additional requirement that this is unique.

Definition 6.2 *A solution ϕ through u is a continuous map $\phi : \mathbf{R} \to Y$ such that $\phi(0) = u$ and $\pi(\phi(\tau), t) = \phi(t + \tau)$ for $t \in \mathbf{R}_+, \tau \in \mathbf{R}$. The range of ϕ is denoted by $\gamma(u)$ and is called the* orbit *through u.*

A set $U \subset Y$ is said to be *forward invariant* if $\gamma^+(U) \subset U$, and *invariant* if $\gamma(U) \subset U$.

The asymptotic behaviour of orbits is next considered. A set U is said to be an *attracting set* for V if $\lim_{t \to \infty} d(v.t, U) = 0 \ (v \in V)$; it is *globally attracting* if $V = Y$. The semiflow is said to be *dissipative* if there is a bounded globally attracting set. U is a *global attractor* if it is compact, invariant and $\lim_{t \to \infty} \overline{d}(V.t, U) = 0$ for all bounded V.

The ω-limit set $\omega(u)$ of u is defined to be the set

$$\omega(u) = \{v : \exists \{t_n\} \text{ with } t_n \to \infty \text{ such that } u.t_n \to v\}.$$

If $\gamma^+(u)$ is relatively compact, then $\omega(u)$ is non-empty, compact, invariant and connected. A concept analogous to that of stable manifold for a smooth dynamical system is that of *stable set* defined as follows for compact invariant U:

$$W^s(U) = \{u : \omega(u) \neq \emptyset, \ \omega(u) \subset U\}.$$

It is natural to expect dissipativity for any realistic biological system, for this is nothing but a reflection of the finiteness of the world. However, the existence of a global attractor is a strong statement as it has the following implications. First, a compactness condition is imposed on the semiflow. Second, there is a full orbit through every point of a global attractor (in

general, only the existence of forward orbits holds for a semiflow). In fact
for a dissipative semiflow arising from a reaction-diffusion system, the
smoothing action referred to above is enough to ensure that semi-orbits
are relatively compact, and the following theorem (Hale & Waltman 1989)
yields a global attractor. Below an operator is said to be compact if it is
continuous and maps bounded sets to relatively compact sets.

Theorem 6.1 *Let Y be complete and suppose that the semiflow is dissipative. Assume that there is a $t_0 \geq 0$ such that $\pi(\cdot, t)$ is compact for
$t > t_0$. Then there is a non-empty global attractor.*

6.2.3   Average Liapunov functions

We are going to assume throughout that the conditions of The-
orem 6.1 hold. We also assume that $Y = Y_0 \cup \partial Y_0$ where Y_0 is open, and
take $Y_0, \partial Y_0$ to be forward invariant. (In the simplest case of ordinary differential equations, $Y = \mathbf{R}_+^n$, and ∂Y_0 is just the boundary of the positive
cone).

Definition 6.3 *The semiflow is said to be* permanent *if there exists a
bounded attracting set A for Y_0 with $\underline{d}(U, \partial Y_0) > 0$.*

We first remark that as far as the theory is concerned, we may treat
reaction-diffusion systems almost as easily as systems of ordinary differential equations. This is because if the conditions of Theorem 6.1 hold,
the smoothing action of the flow makes it possible to find fairly easily
a compact invariant set X; the details will be given elsewhere. With
$S = X \cap \partial Y_0$ (obviously compact), we must show that there is a compact
attracting set M for $X \setminus S$ with $M \cap S = \emptyset$.

In trying to invent a technique for tackling permanence, one first notes
that it is a uniform repelling property of S, and that only the properties
of the semiflow in a neighbourhood of S should be important. Perhaps
the simplest way in which permanence might fail is if there is a point
$P \in S$ such that $W^s(P)$ intersects $X \setminus S$, for then some orbit tends to P.
One might at first be tempted to think that if this possibility is ruled out,
permanence might hold. However, this is not correct as orbits may tend
to S in a more complicated way, for example by spiralling out towards S.
Figure 6.1 shows a two-dimensional section of a three competing species
problem (May & Leonard 1975) which illustrates this point.

There are two main techniques for tackling permanence of which one
is outlined here; both are discussed by Hutson & Schmitt (1990) and by
Waltman (1990). The idea is to weaken radically the requirements on
a Liapunov function (actually for a repeller rather than the more usual
attractor). Recall that with U a neighbourhood of $S, P \in C(U, \mathbf{R}_+)$ is a
Liapunov function if

$$P^{-1}(0) \;=\; S, \tag{6.8}$$

$$P(u.t)/P(u) \;>\; 1 \qquad (t > 0, u \in U \setminus S). \tag{6.9}$$

Thus P increases along orbits, and clearly if this is the case permanence will hold. However, in anything but the simplest situations it turns out to be extremely difficult to find such a function. The broad idea of the central Theorem 6.2 below is to show that it is enough if (6.9) holds

(a) on $\omega(S)$ only after taking a suitable limit,

(b) for some t (rather than all $t > 0$).

These conditions are very much easier to satisfy in many examples, and crucially $\omega(S)$ in (a) may be a much smaller set than S. P below is called an average Liapunov function.

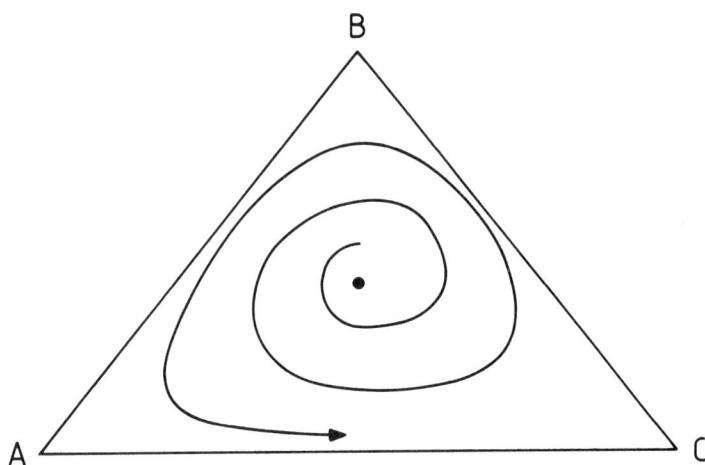

Figure 6.1: $W^s(P)$ does not intersect the interior for any P on the boundary of ABC, but permanence does not hold.

Theorem 6.2 *Let S, X be compact and S have empty interior. Assume that S and $X \backslash S$ are forward invariant. Let $P : X \backslash S \to \mathbf{R}_+$ be continuous, strictly positive and bounded. For $u \in X$ define*

$$\alpha(t, u) = \liminf_{\substack{v \to u \\ v \in X \backslash S}} P(v.t)/P(v),$$

and suppose that

$$\sup_{t>0} \alpha(t, u) > \begin{cases} 1, & u \in \omega(S), \\ 0, & u \in S. \end{cases} \tag{6.10}$$

Then there is a compact forward invariant set M with $\underline{d}(M, S) > 0$ such that every semi-orbit with initial value in $X \setminus S$ eventually enters M.

6.2.4   Applications

For illustration an example is presented of a three species 1 predator – 2 prey system of ordinary differential equations, and it will then be shown that the same example but with diffusion added may be tackled with little extra difficulty.

We consider then the system (6.1) and first show how to deal with condition (6.10). The calculation is a fairly standard one in the context of Liapunov function theory, and as usual 'dot' will denote differentiation along an orbit. With P smooth, for $v \in X \backslash S$,

$$\begin{aligned}
\alpha(t, v) &= P(v.t)/P(v) \\
&= \exp \left\{ \int_0^t \frac{d}{ds} \log P(v.s) ds \right\} \\
&= \exp \left\{ \int_0^t \dot{P}(v.s)/P(v.s) ds \right\} \\
&= \exp \left\{ \int_0^t \psi(v.s) ds \right\}
\end{aligned}$$

where

$$\psi(u) = \dot{P}(u)/P(u) = <\nabla P, g>/P, \tag{6.11}$$

and $< \cdot, \cdot >$ is the scalar product on $\mathbf{R}^n$. In examples ψ usually has a continuous extension to X, and for simplicity we assume this here.

Theorem 6.3 *Assume that $P \in C^1(X \setminus S, \mathbf{R}_+)$, and suppose that ψ is bounded below with a continuous extension to X. Then the system (6.1) is permanent if*

$$\sup_{t>0} \int_0^t \psi(u.s) ds > 0 \quad (u \in \omega(S)). \tag{6.12}$$

As an application consider the Lotka-Volterra system

$$\dot{u}_1 = u_1(a_1 - \epsilon_{11}u_1 - \epsilon_{12}u_2 - \alpha_1 u_3),$$
$$\dot{u}_2 = u_2(a_2 - \epsilon_{21}u_1 - \epsilon_{22}u_2 - \alpha_2 u_3), \qquad (6.13)$$
$$\dot{u}_3 = u_3(-c + \beta_1 u_1 + \beta_2 u_2 - \gamma u_3),$$

where u_1, u_2 represent prey densities and u_3 the predator density. It is assumed that all the parameters are strictly positive. Dissipativity holds, as may easily be shown, so X is compact and $S = X \cap \partial \mathbf{R}_+^n$ is compact. Now on S, (6.13) reduces to three two-dimensional subsystems, and it is standard that the ω-limit sets of S are just equilibria. We make the following simple choice of average Liapunov function

$$P = \prod_{i=1}^{3} u_i^{\alpha_i},$$

where $\alpha_i > 0$ are to be chosen later. Since $\omega(S)$ are just equilibria, (6.12) holds if $\psi > 0$ at these. A simple calculation then shows that it is enough if there exist $\alpha_1, \alpha_2, \alpha_3 > 0$ such that

$$\sum_{i=1}^{3} \alpha_i f_i(u) > 0$$

at the equilibria. There is now nothing more to do — except for some tedious algebra as there are several equilibria. It is found that the system is permanent if there is an interior equilibrium, the determinant of coefficients $D > 0$, and one (or both) of $a_1 \epsilon_{22} > a_2 \epsilon_{12}$, $a_2 \epsilon_{11} > a_1 \epsilon_{21}$ holds. Furthermore one can show that, modulo equalities, these conditions are necessary and sufficient. One may add that numerical evidence suggests that a stable Hopf bifurcation occurs, and there is a globally stable limit cycle which breaks up into a strange attractor while permanence continues to hold. On the other hand there may be an asymptotically stable interior equilibrium, with the system not permanent. For further details see (Hutson & Vickers 1983), and (Hofbauer & Sigmund 1988) for general three-dimensional Lotka-Volterra systems.

Let us consider what more must be done if diffusion is added. We thus examine the system (6.4) with the f_i as above in (6.13). The most convenient setting for the analysis is $C(\Omega)$ with the L_∞ norm. It is fairly easy to establish dissipativity, but if for example $\gamma = 0$ it is much harder to do this (see Hutson & Schmitt 1990). For general Lotka-Volterra systems or more general systems, the proof of dissipativity may present considerable difficulty.

On the face of it there are severe technical difficulties in dealing with systems of reaction-diffusion equations. However, the theory has advanced so much that, with judicious use of say (Mora 1983) and (Henry 1981), it is easy to show that the conditions of Theorem 6.1 are met. We are thus left with doing some routine integration. We define

$$P(u) = \exp\left\{ \int_\Omega \log[\phi(u(x))]dx \right\} \quad (u \in X \setminus S),$$

where ϕ is taken to be the right-hand side of the expression defining P above. Use of the partial differential equations (6.4) and the divergence theorem then yields

$$\alpha(t, u) = \exp\left\{ \int_0^t ds \int_\Omega [\psi(u(x,s)) + Q(u(x,s))]dx \right\}$$

where

$$\psi(u) = \phi^{-1} \sum_{i=1}^n \frac{\partial \phi}{\partial u_i} u_i f_i(u),$$

$$Q(u) = -\sum_{i=1}^n \sum_{j=1}^m \mu_i \frac{\partial}{\partial x_j} \left(\phi^{-1} \frac{\partial \phi}{\partial u_i} \right) \frac{\partial u_i}{\partial x_j}.$$

Typically, in the switch to partial differential equations, there is an additional term Q here, and essentially this must be positive for the method to work. Fortunately this is the case with ϕ chosen as above.

The final problem is to locate the ω-limit sets of the flow on the boundary (taken to be the set where $u_i(x) = 0$ for some i and all $x \in \Omega$). This is now a harder problem, but again for the particular system above, standard Liapunov function techniques may be applied to show that the ω-limit sets consist just of spatially constant equilibria. We thus obtain finally the result that permanence holds under precisely the same conditions as for the original system of ordinary differential equations (6.1). For further details concerning the reaction-diffusion system, see (Hutson & Schmitt 1990).

6.2.5   Conclusions

Broadly there is a good chance of success with tackling permanence for n-species systems if the ω-limit sets of all the $(n-1)$-dimensional subsystems obtained by setting $u_i = 0$ in turn are known. This is true for a wide variety of models ranging from difference equations through differential equations, to differential-delay systems. For systems of ordinary differential equations, this will normally restrict us to three dimensions, and clearly matters will be at least as difficult for reaction-diffusion systems. Returning to the discussion in section 6.1.3, progress in this context can be claimed if a good range of three species problems can be tackled, which is indeed the case with the technique under discussion.

However, there is one class of system where a great deal more progress is possible, and that is with equations of Lotka-Volterra type, see (Hofbauer & Sigmund 1988), and (Hofbauer *et al.* 1987) for the difference equation analogue. The key is the attractive averaging property of Volterra (see section 6.1.3). When used in combination with Theorem 6.3, it turns out that even if the ω-limit sets are unknown (and perhaps very complicated), it is enough to impose the central inequality at the equilibrium only. Based on this idea, four species problems have been tackled (see Kirlinger 1986, 1988) and indeed even some problems of higher dimension.

The technique described does enable considerable progress to be made with the basic issue of coexistence. There is, only too obviously though, a very big gap between the questions that mathematics can currently answer in this area, and those that biologists commonly ask. There is thus plenty of room for innovation here.

6.3   Patterns and travelling waves

6.3.1   Introduction

The previous section was concerned with perhaps the simplest question about a community of species, and that is whether the community survives in the long term. We now turn to two problem areas for models with diffusion where the questions involve a more detailed knowledge of the behaviour of the species. Of course it is then to be expected that theoretical progress will be difficult except in a rather restricted range of cases.

The first area concerns the possible modelling of pattern formation by reaction-diffusion systems (a pattern being a stable stationary spatially inhomogeneous solution). A key contribution, and indeed a very perceptive one, was made by Alan Turing whilst at Manchester. Diffusion is generally thought of as a smoothing or stabilising process, and many would expect that a stable equilibrium of a reaction system would not be disturbed by the introduction of diffusion. Turing observed that under certain circumstances diffusion *can* destabilise a stable equilibrium, and that a pattern can be formed.

The second topic concerns so called travelling wave (or travelling front) solutions. The idea is to model large (usually one dimensional) spatial domains by infinite domains, and to look for solutions which travel without change of shape. For one species quite a lot of progress can be made by fairly elementary arguments. For two or more species even establishing the existence of such solutions presents a very interesting and challenging problem. Mathematically the basic question is whether there are orbits of certain ordinary differential equations joining equilibria in a high dimensional (≥ 4) space. Techniques from such areas as algebraic topology which can resolve the difficult geometrical questions appear to be called for.

6.3.2   Turing instability

The simplest model where this can occur is based on the pair of equations

$$\frac{\partial u_i}{\partial t} = u_i f_i(u) + \mu_i \frac{\partial^2 u_i}{\partial x^2} \qquad (i = 1, 2), \qquad (6.14)$$

with spatial domain **R**. Suppose that the constant $\bar{u}$ is an equilibrium of the reaction system, and linearize about $\bar{u}$, obtaining the system

$$\frac{\partial u_i}{\partial t} = (Au)_i + \mu_i \frac{\partial^2 u_i}{\partial x^2},$$

where A is a constant 2×2 matrix. The following conditions are imposed on the system.

(i) $\bar{u}$ is asymptotically stable for the reaction system. Thus the eigenvalues of A have negative real parts and

$$a_{11} + a_{22} < 0, \qquad (a)$$

$$a_{11}a_{22} - a_{12}a_{21} > 0. \qquad (b)$$

(ii) $a_{11} > 0$, $a_{22} < 0$ and $a_{12} < 0$, $a_{21} > 0$. We comment on this below.

In order to test whether $\bar{u}$ remains stable when diffusion is included, $u = \alpha e^{\lambda t} \cos kx$ is substituted into (6.14), and it is found that there is a nontrivial solution if and only if

$$\begin{vmatrix} \lambda - a_{11} + k^2 \mu_1 & -a_{12} \\ -a_{21} & \lambda - a_{22} + k^2 \mu_2 \end{vmatrix} = 0.$$

The solutions λ have negative real parts if and only if

$$a_{11} + a_{22} - (\mu_1 + \mu_2)k^2 < 0, \qquad (c)$$

$$H(k) \stackrel{\text{def}}{=} \mu_1 \mu_2 k^4 - (\mu_1 a_{22} + \mu_2 a_{11})k^2 + \det A > 0. \qquad (d)$$

Since (a) obviously implies (c), for an instability we require $H(k) < 0$, which is possible if

$$\mu_1 a_{22} + \mu_2 a_{11} > 2(\mu_1 \mu_2 \det A)^{\frac{1}{2}}. \qquad (6.15)$$

One first notes that (6.15) cannot hold if $\mu_1 = \mu_2$, for this requires $a_{11} + a_{22} > 0$ which contradicts (a). However, with $\mu_2 = 1$ and very small μ_1, (6.15) will be satisfied. Then for appropriate k there will be an eigenvalue λ with positive real part, and the solution of (6.14) will grow. The analysis can obviously be extended fairly easily to a finite space domain by appropriate choice of k.

This is the basic mechanism of a Turing instability which is caused by having 'very unequal' diffusion rates. A key assumption is (ii) above, which assures that the pair of species is of 'activator-inhibitor' type. It is not, it might be thought, easy to see why this assumption leads to an instability, and so plausibly to a pattern; for an explanation see (Edelstein-Keshet 1988, §11.6). The difficulty of seeing through this problem is a not inconsiderable tribute to Turing's intuition.

This mechanism has been much exploited. A particularly attractive application is to the modelling of animal coat patterns, see (Murray 1988). (Meinhardt 1982) and (Murray 1989) are strongly recommended for the background and (Edelstein-Keshet 1988) is a nice introductory account. For more of the mathematical details, (Fife 1979) and (Britton 1986) may be consulted.

We conclude this brief introduction with a few remarks on the mathematics. The governing equations are taken to be (6.4)–(6.6) with space domain finite. For a wide variety of systems local bifurcation theory may be employed to show that near an eigenvalue a (stable) pattern emerges. A typical example of such an approach is presented in (Caristi *et al.* 1990), the context being ecological. It is a much harder matter to give a global theory, and the stability theory for patterns is particularly difficult. To get an idea of the problem, the reader may consult a numerical study by Eilbeck (1989) showing how very complex the bifurcation picture can be for a perfectly reasonable pair of equations. Obviously there is much room for further investigation, both in tackling new models, and in investigating the sort of question just raised.

6.3.3 *Travelling waves*

To fix ideas let us consider first a one-species case. The governing equation is

$$\frac{\partial u}{\partial t} = g(u) + \mu \frac{\partial^2 u}{\partial x^2}, \tag{6.16}$$

and a solution of the form $u(x - ct)$ is required. With $z = x - ct$ as a new variable, substitution in (6.16) gives

$$\mu \ddot{u} + c\dot{u} + g(u) = 0 \tag{6.17}$$

where 'dot' $= d/dz$. The following two basic cases have been much studied, see (Fife 1979) and (Britton 1986).

(i) $g(0) = g(1) = 0$, $g(u) > 0$ ($u \in (0,1)$). With $g(u) = u(1 - u)$ this is the case originally studied by Fisher in a genetics context.

(ii) $g(0) = g(\alpha) = g(1) = 0$ for some $\alpha \in (0,1)$ and $g(u) < 0$ ($u \in (0, \alpha)$), $g(u) > 0$ ($u \in (\alpha, 1)$). A typical example is $g(u) = u(u - \alpha)(1 - u)$. This case is associated initially with Kanel, Aronson and Weinberger, and Fife and McLeod. It is known as the bistable case as the reaction system has two stable equilibria, 0 and 1.

In both cases the aim is to find a solution such that $u(-\infty) = 0$, $u(\infty) = 1$, which is sometimes called a travelling front because u approaches different limits as $z \to \pm\infty$. With (6.17) rewritten in the phase plane as

$$\dot{u} = p,$$
$$\mu\dot{p} = -cp - g(u),$$

the problem is then to find a heteroclinic connection of the equilibria $(0,0)$ to $(1,0)$. Local analysis shows that one is looking for a node-saddle connection in case (i) and a saddle-saddle connection in case (ii). In (i) it can be shown fairly easily that there is a critical number $c^* > 0$ such that a monotone wave (that is $\dot{u} > 0$) exists for all $c \le -c^*$, but not for $c > -c^*$. The second case is harder, intuitively because it is more difficult to connect two saddles, and is the prototype for the cases we principally discuss. The unstable manifold of $(0,0)$ and the stable manifold of $(1,0)$ are both one-dimensional. Since there is one parameter c at our disposal, we expect to be able to find a connection for a unique value of c.

A suggestive argument for problems of this nature may be based on the diagrams in Figure 6.2.

It is fairly clear by a continuity argument that for some intermediate value of c there is a connection of A to B, and it is quite easy to make this argument rigorous, confirming the above assertion in particular. The argument is basically a 'shooting' argument. One may note though that because of the extra 'space' in higher dimensions it will be a great deal less easy to decide whether saddle-saddle connections exist for analogous problems.

Fairly recently attention has shifted to the question of finding travelling waves for two species, and this presents a problem which is an order of magnitude harder. Although there are other approaches, see for example (Dunbar 1984), one of the most promising lines of attack was provided by the work of Conley, a typical result being given in (Conley & Gardner 1984). Unfortunately the techniques are difficult to use. In (Mischaikow & Hutson 1990) we show that at least for a range of problems, an argument in the spirit of those of Conley can be simplified to the extent that it becomes almost a routine calculation. Below we attempt to give a flavour of the ideas used. It is helpful here to bear classical degree theory in mind as general background. For illustration consider a pair of mutualist (sometimes known as symbiotic) species. The corresponding phase plane is shown in Figure 6.3, where O and Q are sinks and P is a saddle.

A solution is required of the following four dimensional problem

$$\dot{u} = p,$$
$$\dot{v} = q,$$
$$\mu\dot{p} = -cp - f(u, v),$$
$$\nu\dot{q} = -cq - g(u, v),$$

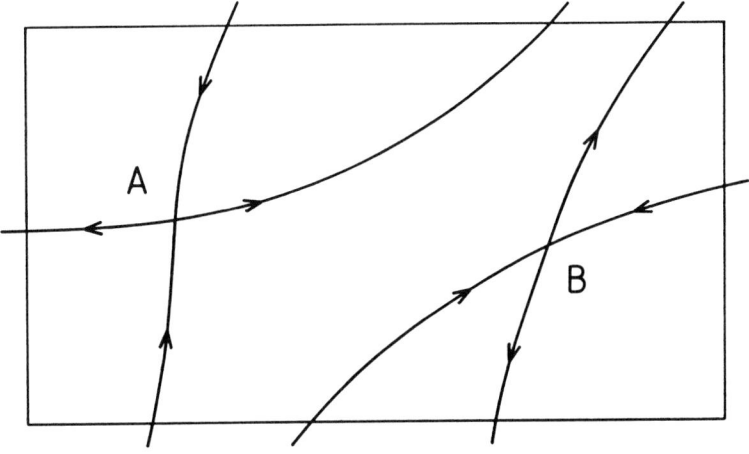

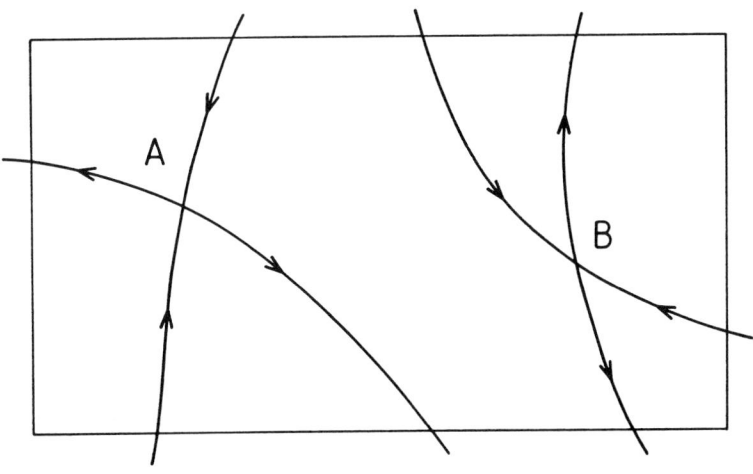

Figure 6.2: The upper and lower figures are phase portraits for large and small c respectively.

joining $(0,0,0,0)$ at $z = -\infty$ with $(1,1,0,0)$ at $z = \infty$.

One's first thought might be to use the very powerful techniques of degree theory, but this program is difficult to carry through as it stands as the domain is not compact; this objection may be circumvented although not very easily, see (Gardner 1982). A somewhat analogous idea may be used though. First one selects an 'isolating neighbourhood' for the flow (the boxes in Figure 6.2 would be an example) analogous to the basic open set used in degree theory, and constructs a topological index based on the flow and how it exits from the isolating neighbourhood. In the above example the index must distinguish the two situations shown. Second, homotope the original equations to a symmetric pair (effectively yielding the one species case), or at least to a pair where the index can be calculated. Let us assume that as in the figure, this is different for two values of c. Of course no solution must 'escape' the isolating neighbourhood; as with degree theory proofs this can require a very careful discussion based on the differential equations. Finally, as the index is invariant under homotopy, the original system must have different indices for the two values of c, and so a solution for some c.

These arguments can also be used for n-mutualists and for a pair of competing species. Of course existence is by no means the only problem of biological interest. For example one wishes to know the sign of c

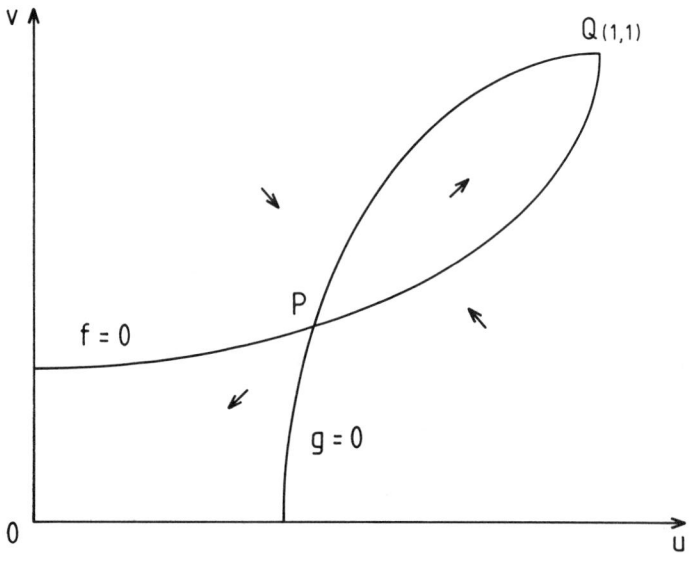

Figure 6.3: *The reaction phase plane for a pair of mutualists*

as this determines whether species grow or die out, a matter of obvious importance in applications. For much interesting background, see (Murray 1989). The known results answer only a very small fraction of the questions in this area, and there remains a great deal of work to be done theoretically and computationally.

6.4   References

Arthur, W. 1990 Positive feedback in the economy. *Sci. Amer.* **262**, 92–99.

Bhatia, N. & Szegö, G. 1970 *Stability theory of dynamical systems.* Berlin: Springer.

Britton, N.F. 1986 *Reaction diffusion equations and their applications.* London: Academic Press.

Burton, T. & Hutson, V. 1989 Repellers in systems with infinite delay. *J. Math. Anal. Appl.* **137**, 240–263.

Caristi, G., Rybakowski, K. & Wessolek, T. 1990 Persistence and spatial patterns in a one-predator-two-prey Lotka-Volterra model with diffusion. *SIAM J. Appl. Math.*, to appear.

Conley, C. & Gardner, R. 1984 An application of the generalized Morse index to travelling wave solutions of a competitive reaction-diffusion model. *Indiana. J. Math.* **33**, 319–345.

Dunbar, S. 1984 Travelling wave solutions of diffusive Lotka-Volterra equations: a heteroclinic connection in R^4. *Trans. Amer. Math. Soc.* **286**, 557–594.

Edelstein-Keshet, L. 1988 *Mathematical models in biology.* New York: Random House.

Eilbeck, C. 1989 Theoretical biology: epigenetic and evolutionary order. In *Waddington memorial conference* (ed. B. Goodwin & P. Saunders), pp. 31–41.

Fife, P. 1979 *Mathematical aspects of reacting and diffusing systems.* New York: Springer.

Gardner, R. 1982 Existence and stability of travelling wave solutions of competition models: a degree theoretical approach. *J. Diff. Eqns.* **44**, 343–364.

Hale, J.K. 1988 *Asymptotic behavior of dissipative systems.* Mathematical Surveys & Monographs 25, Amer. Math. Soc.

Hale, J.K. & Waltman, P. 1989 Persistence in infinite dimensional systems. *SIAM J. Math. Anal.* **20**, 388–395.

Henry, D. 1981 *Geometric theory of semilinear parabolic equations.* Lecture Notes in Maths. **840**.

Hofbauer, J., Hutson, V. & Jansen, W. 1987 Coexistence for systems governed by difference equations of Lotka-Volterra type. *J. Math. Biol.* **25**, 553–570.

Hofbauer, J. & Sigmund, K. 1988 *Dynamical systems and the theory of evolution.* Cambridge: University Press.

Hoppensteadt, F. 1982 *Mathematical models in population biology.* Cambridge: University Press.

Hutson, V. & Moran, W. 1982 Persistence of species obeying difference equations. *Math. Biosci.* **15**, 203–213.

Hutson, V. & Schmitt, K. 1990 Permanence in dynamical systems. Submitted to *Dynamics Reported.*

Hutson, V. & Vickers, G. 1983 A criterion for permanent coexistence of species, with an application to a two-prey one-predator system. *Math. Biosci.* **63**, 253–269.

Kirlinger, G. 1986 Permanence in Lotka-Volterra equations: Linked predator-prey systems. *Math. Biosci.* **82**, 165–191.

Kirlinger, G. 1988 Permanence of some ecological systems with several predator and one prey species. *J. Math. Biol.* **26**, 217–232.

Lasota, A. & Mackey, M. 1985 *Probabilistic properties of deterministic systems.* Cambridge: University Press.

Lotka, A. 1956 *Elements of mathematical biology.* New York: Dover.

May, R. & Leonard, W.J. 1975 Nonlinear aspects of competition between three species. *SIAM J. Appl. Math.* **29**, 243–253.

Meinhardt, H. 1982 *Models of biological pattern formation.* New York: Academic Press.

Mischaikow, K. & Hutson, V. 1990 Travelling waves for mutualist species. In preparation.

Mora, X. 1983 Semilinear parabolic problems define semiflows in C^k spaces. *Trans. Amer. Math. Soc.* **278**, 21–55.

Murray, J. 1988 How the leopard gets its spots. *Sci. Amer.* **258**, 80–87.

Murray, J. 1989 *Mathematical biology.* Berlin: Springer.

Okubo, A. 1980 *Diffusion and ecological problems: mathematical models.* Lecture Notes in Biomath. **10**.

Saperstone, S.H. 1981 *Semidynamical systems in infinite dimensional spaces.* New York: Springer.

Schaffer, W., Ellner, S. & Kot, M. 1986 Effects of noise on some dynamical models in ecology. *J. Math. Biol.* **24**, 479–523.

Schuster, P., Sigmund, K. & Wolff, R. 1979 Dynamical systems under constant organisation 3: Cooperative and competitive behaviour of hypercycles. *J. Diff. Eqns.* **32**, 357–368.

Volterra, V. 1931 *Leçons sur la théorie mathématique de la lutte pour la vie.* Paris: Gauthier-Villars.

Waltman, P. 1990 A brief survey of persistence. *Proc. 65th Birthday Conf. for Kenneth Cooke*, to appear.

7

The approximation of free-surface Green functions

J.N. Newman
Massachusetts Institute of Technology

As an aspiring student of naval architecture I was
frustrated by free-surface Green functions until Fritz Ursell
came to my rescue. His analytical expertise was matched by
his patience in teaching me the mathematical subtleties of
hydrodynamics. In the subsequent three decades he has
continued to offer generous assistance whenever an integral
or expansion seemed to present insurmountable difficulties.
It is a great pleasure to participate in this meeting
honouring Professor Ursell.

7.1 Introduction

In boundary-value problems for the velocity potential exterior to
a floating or submerged body, with a linearized free-surface condition and
appropriate radiation condition at infinity, the solution may be derived
from Green's theorem. An alternative approach for simple body shapes
is to use multipole expansions. The fundamental solution in either case
is the source potential which satisfies the free-surface condition, *i.e.* the
free-surface Green function.

The principal free-surface Green functions are collected in analytic form
by Wehausen & Laitone (1960). Typically these are defined by integral
representations, which can be interpreted as transforms in wavenumber
space over the domain of the free surface. Most applications are in the
frequency domain, where the time-dependence is harmonic with a prescribed frequency ω. The corresponding integrands are oscillatory and
include poles at the roots of the plane-wave dispersion relation. Direct
numerical integration is inefficient, particularly if the source and field
point are close to the free surface.

The first useful computations of the simplest two-dimensional free-
surface Green function were presented by Fritz Ursell in his classical paper
on the oscillatory motions of a floating circular cylinder (Ursell, 1949). A
modified multipole technique was devised with 'wave-free' higher-order
multipoles; as a result the only computations required of the free-surface

integral were for the source itself, with the source-point at the origin. The same method was used by Havelock (1955) for the analogous three-dimensional problem of a floating hemisphere.

Today it is routine to solve three-dimensional radiation or diffraction problems with $O(10^3)$ discretized panels on the body surface, involving $O(10^6)$ evaluations of the Green function for each frequency. Including only the computations performed in our group at MIT we estimate that 10^{10} evaluations of the frequency-domain Green function have been made in the past two years. Clearly such a level could not be achieved without using special algorithms to evaluate these functions.

From the computational standpoint there are important analogies between free-surface Green functions and Bessel functions. Both can be defined by integral representations and evaluated by quadratures, with a minimum of programming effort, but the computing time is excessive and the robustness is questionable if the argument is large. A suitable combination of the ascending power series and asymptotic expansion provides better algorithms for evaluating, say, the Bessel function $J_0(x)$, but this strictly analytic approach is not without problems. To compute J_0 with an absolute accuracy of 6 decimals requires ten terms in the ascending series, for $|x| \leq 5$, and eleven terms in the asymptotic expansion for the complementary domain $|x| \geq 5$. One significant digit is lost in the ascending series, hence single-precision arithmetic is marginal. Results with increased accuracy are more difficult to obtain, and on a typical computer with 16-17 decimals accuracy in double-precision arithmetic 4-5 decimals are lost due to cancellation errors near the optimum partition. These difficulties can be overcome, and greater computational efficiency can be achieved, if the analytic expansions are replaced by more effective approximations such as truncated Chebyshev expansions or the equivalent economized polynomials.

In evaluating free-surface Green functions three obstacles stand in the way of more efficient approaches:

1. the functions to be evaluated depend on two or three physical coordinates, and usually on additional parameters, all of which may take a large range of values;
2. the user community is small, and
3. there is no compendium of analytic knowledge analogous to Watson's treatise!

Notwithstanding these obstacles, the application of numerical approximations to free-surface Green functions can lead to substantial improvements in computational efficiency and robustness. As with the simpler Bessel functions, it is vital to separate singularities and persistent oscillations from the slowly-varying residuals or factors. In this respect analytic knowledge is essential, and the most progress has been achieved where we have a comprehensive understanding of the singular features.

My first work in this area was devoted to the three-dimensional frequency-domain Green function, where practical applications abound and the mathematical difficulties are minimal. Emboldened by the success resulting from *ad hoc* combinations of analytic expansions and multivariate polynomial approximations, I considered some aspects of the steady translating Green function, important in the wave-resistance of ships, and also the time-domain impulsive Green function. The results are described in three references (Newman, 1985a,b,1987). Motivated by the need to further reduce computational costs, particularly on vector computers such as the Cray, I have returned to this topic recently with a more heavy-handed reliance on polynomial approximations and less use of analytic expansions except for guidance in the form of the approximations.

Various methods of approximation are available, but I prefer the use of multivariate Chebyshev expansions which lead to economized polynomials with near minimax accuracy or 'equal-ripple' errors. Algorithms for determining the polynomial coefficients, outlined in Newman (1987), are extensions of the procedure for approximating a function of a single real variable by an economized polynomial.

The numerical technique used to derive the polynomial approximations is described in §7.2. In §7.3 a simple periodic array of Rankine sources is analysed to illustrate the utility of the method. In this case a single approximation subdomain is sufficient to complement the more familiar Fourier series. Approximations of the frequency-domain Green functions are described in §§7.4-7.5 for infinite and finite fluid depths, respectively. For these functions the analytic and asymptotic properties are well established, and attention is focussed on more unified and efficient approximations which are suitable for vectorization.

In §§7.6-7.7 the analogous Green functions are considered for impulsive motion in the time domain. Special attention is given to the large-time asymptotics, which are prerequisites to the development of corresponding numerical approximations. In the infinite-depth case the method of steepest descents is utilized in a straightforward manner. The asymptotics are more complicated in a fluid of finite depth, where the front which advances with the limiting group velocity separates the oscillatory waves from an exponentially small disturbance far ahead. The transition between these two regions suggests an asymptotic expansion in terms of Airy integrals using the method of Chester, Friedman & Ursell (1957), as in the analysis of the two-dimensional Cauchy-Poisson problem by Whitham (1974). However this approach is not directly applicable in the three-dimensional case, where we show that the wave front is described to leading order by the *square* of the Airy integral.

In the concluding §7.8 the use of polynomial approximations is compared with other computational approaches based on table look-up and interpolation. Progress is briefly noted on the analysis of ships, where a steady forward velocity is superposed.

7.2   Polynomial approximation

Given a function $F(x, y)$ which is regular in the domain $|x| \leq 1$, $|y| \leq 1$, a Chebyshev expansion may be assumed in the form

$$F = \sum_{m=0}^{\infty} \sum_{n=0}^{\infty} c_{mn} T_m(x) T_n(y) \tag{7.1}$$

where T_m is the Chebyshev polynomial. Since F is regular the sequence of coefficients c_{mn} will converge to zero and the series may be truncated to yield any desired degree of accuracy. The extension to three or more variables is straightforward.

For computational convenience the result can be converted to an ordinary polynomial. Gradients of the Green function may be evaluated using term-by-term differentiation of the polynomials, but extra accuracy is necessary in the approximations to offset the loss inherent in differentiation. Nested multiplication permits the polynomials to be evaluated with one floating-point multiplication and addition per term. Derivatives can be evaluated concurrently, with the same computational cost, by using a simple extension of the nested-multiplication algorithm.

The requirement that F be 'regular' is not strict. Weak singularities such as $x^k \log x$ can be tolerated if k is sufficiently large. This pragmatic consideration is significant if we lack a complete understanding of the singularity, or if a simpler representation is expedient from the computing standpoint. Generally the suitability of F for representation in the form (7.1) can be judged by superficial examination of the coefficients. One learns from experience to determine the types of singularities which have been overlooked, or incorrectly removed, by the resulting sequence of coefficients. Table 7.2 shows a 'good' matrix of Chebyshev coefficients, corresponding to the economized polynomial coefficients $a_{mn}^{(+)}$ in equation (7.8) and Table 7.3. The discrete orthogonality relation is convenient for computing the Chebyshev coefficients. An accurate subroutine is essential for the evaluation of F, ideally with several more decimals precision than is needed in the final approximation. Efficiency is not important, since

n	$m = 0$	$m = 1$	$m = 2$	$m = 3$	$m = 4$
0	0.01207372	0.02464014	0.00025322	0.00000300	*0.00000004*
1	-0.01393953	-0.00178218	-0.00005811	-0.00000146	*-0.00000003*
2	0.00014993	0.00005214	0.00000332	*0.00000014*	0.00000000
3	-0.00000242	-0.00000140	*-0.00000014*	-0.00000001	-0.00000000
4	*0.00000005*	*0.00000004*	0.00000001	0.00000000	0.00000000

Table 7.1: Coefficients c_{mn} of the Chebyshev expansion (7.1) for the source array G_+ defined in (7.7). Coefficients with an absolute value less than 10^{-6} are italicized.

merical integration of the Green functions is appropriate in this context, provided an adaptive method such as Romberg quadrature is employed to assure the required degree of accuracy. When using the discrete orthogonality relation it is necessary to truncate the Chebyshev expansions *a priori* at some appropriate upper limits $m = M$, $n = N$ which exceed the ultimate degrees of truncation based on the economization tolerance. In practice $M = N = 8$ or 16 are convenient choices. A preliminary examination of the convergence will indicate if these are not sufficiently large, in which case it may be more appropriate to subdivide the domain of approximation (or reconsider the singular features of F!) rather than increasing M or N.

The accuracy required for approximations of Green functions is an important practical issue. For applications with integrated surface distributions there is no point in preserving the relative accuracy locally at points where the function is small. The present discussion is in the context of approximations with 6-7 decimals absolute accuracy, the maximum consistent with single-precision arithmetic on most computing systems. Polynomial approximations with greater or lesser accuracy may be derived by the same techniques, and require larger or smaller numbers of coefficients, respectively.

It might be argued that 6-7 decimals are unnecessary for engineering applications, where precision greater than 2-3 decimals in the final results usually is irrelevant. On the other hand, since the gradient of the Green function is required for the coefficient matrix, and the condition number of the resulting linear system is frequently $O(100)$, 6-7 decimals are not excessive. Furthermore, validation of programs and numerical methods depends on convergence tests and comparisons with benchmarks, which should be made to a higher degree of accuracy than is required in engineering practice.

The algorithm for truncating the Chebyshev expansion must be considered together with the desired accuracy. Since $|T_m| \leq 1$ the contribution from each separate term is bounded by $|c_{mn}|$. The simplest choice is to ignore any coefficient with magnitude less than a prescribed tolerance. A more conservative criterion is to sum the absolute values of all coefficients backwards from large upper limits, and ignore all terms if the sum itself is less than the tolerance. In one dimension this process is unique, but in multiple dimensions there is some flexibility. In Table 7.2, for example, the absolute sum of the italicized terms is 4.6×10^{-7}. Setting all of these terms equal to zero will therefore result in a maximum economization error of the same amount. On the other hand, if a tolerance of 4.0×10^{-7} is imposed, either the coefficient c_{23} or the coefficient c_{32} could be removed, but not both. This flexibility in truncation can be exploited to preserve the form of the polynomial approximations in different subdomains, *i.e.* the same number of coefficients is retained in each column and each row of the truncated matrix. This is an important practical feature from the programming standpoint since the same code may be used for every sub-domain if the coefficients are stored in a common array.

7.3   Array of Rankine image sources

In two limiting cases the free-surface boundary condition reduces to a homogeneous Neumann or Dirichlet condition. These correspond respectively to $\omega = 0, \infty$ in the frequency domain, and $t = \infty, 0$ for impulsive motion in the time domain. For a fluid of infinite depth the solution is simply a Rankine source $1/r$ together with its image of the same or opposite sign, above the plane of the free surface. For finite depth the solution can be constructed from a pair of periodic source arrays

$$
G_{\pm}(R, z) = \frac{1}{\sqrt{R^2 + z^2}}
$$
$$
+ \sum_{\substack{n=-\infty \\ n \neq 0}}^{\infty} (\pm)^n \left\{ \frac{1}{\sqrt{R^2 + (z + 2n)^2}} - \frac{1}{2|n|} \right\} \tag{7.2}
$$

where the two alternatives correspond respectively to Neumann and Dirichlet conditions on the boundary $z = 1$. To secure convergence in the former case a constant has been subtracted from each term. The cylindrical coordinates (R, z) are defined with the array on the z-axis and the normalized spacing equal to two. The function G_- can be derived by forming the difference between two functions G_+, with an appropriate shift of the origin, but it is useful to treat these alternative cases separately.

Fourier series are easily derived, as in Gradshteyn & Ryzhik (1965), §8.526. The resulting expressions are

$$
G_+ = 2 \sum_{m=1}^{\infty} \cos(m\pi z) K_0(m\pi R) - \gamma - \log(\tfrac{1}{4}R) \tag{7.3}
$$

$$
G_- = 2 \sum_{m=0}^{\infty} \cos((m + \tfrac{1}{2})\pi z) K_0((m + \tfrac{1}{2})\pi R) + \log 2 \tag{7.4}
$$

Here K_0 is the modified Hankel function and $\gamma = 0.577...$ is Euler's constant. These series converge rapidly, except for small values of R. If $R > 1$ a maximum of six terms is sufficient to give 7-8 decimals accuracy.

The objective here is to complement (7.3, 7.4) with alternative algorithms which are effective when $R \leq 1$. Since both functions are periodic in z, with obvious symmetry or anti-symmetry relations about the planes $z = 0, \pm 1, \pm 2, ...$, it may be assumed hereafter that $0 \leq z \leq 1$.

If the Hankel transform for the inverse square-root is substituted for each term in (7.2) the series can be summed to give the integral representations

$$
G_{\pm} = \frac{1}{\sqrt{R^2 + z^2}} \pm \int_0^{\infty} I_{\pm}(k, z, R) \, dk, \tag{7.5}
$$

where

$$I_\pm(k, z, R) = e^{-k}\binom{\operatorname{csch} k}{\operatorname{sech} k}\big[\cosh(kz)J_0(kR) - 1\big].$$

These integrals converge if $|z| < 2$. For sufficiently small values of R and z the integrand can be expanded in even powers of these coordinates, and integrated term-by-term. After using integral representations for the (Riemann) zeta function ζ, it follows that

$$
\begin{aligned}
G_\pm &= [R^2 + z^2]^{-\frac{1}{2}} \\
&+ \sum_{m=0}^\infty \sum_{n=0}^\infty (-)^m (1 - \delta_{m+n}^0)\frac{(2m + 2n)!}{(m!)^2(2n)!} \\
&\times \left(\frac{1}{1 - 2^{-2m-2n}}\right)\zeta(2m + 2n + 1)(\tfrac{1}{4}R)^{2m}(\tfrac{1}{2}z)^{2n}
\end{aligned}
\tag{7.6}
$$

An equivalent result for G_+ has been derived by Breit (1990) using Taylor expansions.

As the basis for more efficient computational algorithms, the adjacent singularities in (7.2) corresponding to the terms $n = \pm 1$ are subtracted from the integral representation to accelerate the convergence as $k \to \infty$:

$$
\begin{aligned}
G_\pm &= \frac{1}{\sqrt{R^2 + z^2}} \pm \frac{1}{\sqrt{R^2 + (z + 2)^2}}\frac{1}{\sqrt{R^2 + (z - 2)^2}} \mp 1 \\
&\pm \int_0^\infty I_\pm(k, z, R) - 2e^{-2k}\cosh(kz)J_0(kR)\, dk.
\end{aligned}
\tag{7.7}
$$

In the domain $(0 \leq R \leq 1,\ 0 \leq z \leq 1)$ these integrals define regular functions, which can be approximated by economized polynomials. The results take the form

$$
\begin{aligned}
G_\pm &\simeq \frac{1}{\sqrt{R^2 + z^2}} \pm \frac{1}{\sqrt{R^2 + (z + 2)^2}} \pm \frac{1}{\sqrt{R^2 + (z - 2)^2}} \mp 1 \\
&+ \sum_{m,n} a_{mn}^{(\pm)} R^{2m} z^{2n}
\end{aligned}
\tag{7.8}
$$

where the coefficients $a_{mn}^{(\pm)}$ are given in Tables 7.3 and 7.3. These coefficients have been obtained in the manner outlined in §2, based on Chebyshev expansion of the functions defined by the integrals in (7.7), and neglect of the Chebyshev coeficients with magnitude smaller than 10^{-9}. The resulting polynomial approximations are accurate to about 8 decimals. (Table 7.2 shows the Chebyshev coeficients corresponding to a_{mn}^+.)

The utility of this approach is that the slowly-convergent series (7.2) (excluding the three dominant terms $n = -1, 0, 1$) are approximated by polynomials which can be evaluated with 21 floating-point multiplications/additions. Similar approximations can be derived directly from (7.5) or (7.6), without subtracting the adjacent singularities in (7.7), but the convergence of the Chebyshev expansions is reduced and more terms are required to achieve the same accuracy.

7.4   Frequency domain – infinite depth

If the time-dependence is harmonic, with frequency ω, the Green function is defined by the expression

$$
\begin{aligned}
G &= 1/r + 1/r_1 + \int_0^\infty \frac{k+K}{k-K} e^{kz} J_0(kR)\,dk \\
&= 1/r + K F(KR, |Kz|) - 2\pi i K e^{Kz} J_0(KR) \qquad (7.9)
\end{aligned}
$$

(Wehausen & Laitone, 1960, eq. 13.17). Here $1/r_1$ is the potential of the image source above the free surface, $K = \omega^2/g$ is the wavenumber with g the gravitational acceleration constant, and (R, z) are cylindrical polar coordinates with z positive upwards and the origin at the image source point. The time-dependent factor $\exp(i\omega t)$ is assumed, and the contour is deformed above the pole in accordance with the radiation condition. The essential task is to evaluate the function $F(X, Y)$ throughout the quadrant $(0 \le X < \infty,\ 0 \le Y < \infty)$, excluding the singular point at the origin.

Newman (1985a) describes a set of comprehensive but disjoint algorithms for this purpose. Different analytic expansions are used in four subregions around the perimeter of the (X, Y) quadrant, complemented by economized polynomials in the interior region where the analytic expansions are not efficient. The resulting subroutine is robust, and efficient on a serial computer, but the variety of diverse algorithms and associated branches precludes effective use on a vector machine such as the Cray.

n	$m = 0$	$m = 1$	$m = 2$	$m = 3$	$m = 4$
0	0.00000000	-0.02525711	0.00086546	-0.00004063	0.00000193
1	0.05051418	-0.00692380	0.00073292	-0.00006636	0.00000398
2	0.00230838	-0.00097875	0.00020597	-0.00003333	0.00000524
3	0.00012934	-0.00010879	0.00003965	-0.00000891	
4	0.00000913	-0.00001270	0.00000466		

Table 7.2: Coefficients $a_{mn}^{(+)}$.

n	$m = 0$	$m = 1$	$m = 2$	$m = 3$	$m = 4$
0	0.00000000	0.01230716	-0.00065341	0.00003603	-0.00000182
1	-0.02461428	0.00522739	-0.00065004	0.00006286	-0.00000386
2	-0.00174290	0.00086822	-0.00019543	0.00003259	-0.00000519
3	-0.00011463	0.00010319	-0.00003877	0.00000882	
4	-0.00000870	0.00001243	-0.00000460		

Table 7.3: Coefficients $a_{mn}^{(-)}$.

Adopting a more systematic numerical approach, we use the analytic expansions only as guides for removal or suppression of the singular or oscillatory features of $F(X,Y)$, and approximate the remainder in all cases with economized polynomials. For this purpose the quadrant is subdivided into four rectangular domains with partitions at $X = 3$ and $Y = 4$. From the expansions given by Newman (1985a) four corresponding approximations can be inferred:

Region 1 $(0 \leq X \leq 3,\ 0 \leq Y \leq 4)$:

$$
\begin{aligned}
F(X,Y) &= -2e^{-Y}J_0(X)\log(R+Y) \\
&- e^{-Y}\left[\pi Y_0(X) - 2J_0(X)\log(x)\right] \\
&+ e^{-Y}Rf(X^2,Y)
\end{aligned}
\tag{7.10}
$$

Region 2 $(3 \leq X < \infty,\ 0 \leq Y \leq 4)$:

$$
F(X,Y) = -2\pi e^{-Y}Y_0(X) + f(X^{-1},Y)
\tag{7.11}
$$

Region 3 $(0 \leq X \leq 3,\ 4 \leq Y < \infty)$:

$$
F(X,Y) = f(X^2,Y^{-1})
\tag{7.12}
$$

Region 4 $(3 \leq X < \infty,\ 4 \leq Y < \infty)$:

$$
\begin{aligned}
F(X,Y) &= -2\pi e^{-Y}Y_0(X) + \sum_{n=0}^{2} n!P_n(\cos\theta)R^{-(n+1)} \\
&+ f(X^{-1},Y^{-1})
\end{aligned}
\tag{7.13}
$$

Here J_0, Y_0 denote the Bessel functions, R, θ are spherical coordinates with $Y = R\cos\theta$, and $P_n(\cos\theta)$ is the Legendre polynomial. In each region f denotes a (different) residual function of the two indicated variables, to be approximated by an economized polynomial.

The success of this approach depends on the extent to which the residual function f is slowly-varying. In Region 1 the logarithmic singularity is represented exactly, and the remainder f is strictly regular. Similarly in Region 3, the asymptotic behaviour of the integral in (7.9) can be used to confirm the assumed form. However in Regions 2 and 4 the extra terms are somewhat *ad hoc*. A more correct form would replace $2Y_0$ by the sum $Y_0 + \mathbf{H}_0$ where $\mathbf{H}_0$ is the Struve function, but since the difference $Y_0 - \mathbf{H}_0$ is algebraic in X^{-1} it is unnecessary to evaluate $\mathbf{H}_0$. Including the three-term spherical-harmonic series in Region 4 reduces the variation of the residual function.

In principle a single polynomial approximation is feasible in each domain, but to avoid excessive numbers of terms it is necessary to introduce additional partitions. Numerical experimentation indicates that a total of 48 subdomains are effective, separated by partitions at $X = 3, 3.75, 5, 7.5, 15$ and $Y = 1, 2, 3, 4, 5\frac{1}{3}, 8, 16$. A total of 31 terms in each economized double-polynomial is suitable to yield an absolute accuracy of 6 decimals for each subdomain. Aside from the relatively simple extra terms required in Regions 1,2, and 4, the computation of F is reduced to the evaluation of this polynomial. A total of $48 \times 31 = 1488$ polynomial coefficients must be stored in the program.

From the programming standpoint the complications of a large number of subdomains can be minimized by using polynomials of the same degree, and storing the coefficients sequentially in a large array. A single block of nested-multiplication code can then be used for all cases. This code can be vectorized, thus permitting the simultaneous evaluation of a large number of Green functions in one subroutine call; on a Cray this procedure reduces the computational cost of each Green-function evaluation by a factor of 5-10.

7.5   Frequency domain – finite depth

For a fluid of constant depth h with vanishing normal velocity on the bottom, the extension of (7.9) is given by

$$
\begin{aligned}
G & = \frac{1}{r} + \frac{1}{r_2} \\
& + \ 2 \int_0^\infty \frac{k + K}{k \sinh(kh) - K \cosh(kh)} e^{-kh} \mathcal{G}(\zeta, z, R) \, dk, \quad (7.14)
\end{aligned}
$$

where

$$
\mathcal{G}(y, z, R) = \cosh(k(y + h)) \cosh(k(z + h)) J_0(kR),
$$

(Wehausen & Laitone, 1960, eq. 13.18). Here (R, z) is a cylindrical coordinate system with vertical axis, the fluid domain is $(-h \leq z \leq 0)$, the source point is at $(0, \zeta)$, and $1/r_2$ is the potential of the image source beneath the bottom. As in the infinite-depth case the appropriate contour of integration is above the pole.

An eigenfunction expansion analogous to (7.3, 7.4) (Wehausen and Laitone, 1960, eq. 13.19) can be used effectively for computations unless the ratio R/h is small. The procedure adopted by Newman (1985a) is based on a partition at $R/h = 1/2$. For $R/h \geq 1/2$ a maximum of 12 terms in the eigenfunction expansion gives an accuracy of 6 decimals. In the complementary domain $R/h \leq 1/2$ the hyperbolic addition theorem is used to express the Green function, initially dependent upon four nondimensional parameters $(R/h, z/h, \zeta/h, Kh)$, as the sum of two auxiliary functions L which depend on only three parameters $(X = R/h, V = |z/h \pm \zeta/h|, Kh)$. (The function L involves an integral similar to that in (9) but with $\zeta = -h$; the same decomposition is used in §7.7.) The convergence of the integral representation for L can be accelerated by subtracting two infinite-depth integrals of the form (7.9). The difference is a regular function of the three parameters which can be approximated by *triple* economized polynomials (except in the limit $Kh \to 0$ where a logarithmic singularity must be accounted for). With three subdomains in Kh, 6 decimals accuracy requires about 300 terms, but in the usual applications involving many values of the source and field points for the same frequency and depth, the polynomials in Kh can be evaluated at the outset, leaving only 33 terms for the polynomials in terms of (X, V). The resulting algorithm for G requires two evaluations of this polynomial, and four evaluations of the infinite-depth Green function.

As in the infinite-depth case, recent improvements have been made in these algorithms. The eigenfunction expansion is used only in the domain $X \geq 1$, thereby halving the maximum number of terms required. In the domain $(0 \leq X \leq 1)$ two evaluations of the auxiliary function $L(X, V, Kh)$ are used, as before, but the domain of the vertical coordinate is subdivided at $V = 1$ and the infinite-depth limit of the integrand is subtracted only when $V \leq 1$. To cover the extended domain $(0 \leq X \leq 1)$ more partitions are required in the parameter Kh, and the total number of stored coefficients increases to about 8000. For a fixed value of Kh the Green function is evaluated from two polynomials with 32 terms in each, and from only one or two evaluations of the infinite-depth function. In addition to a modest reduction in serial computing time, the resulting algorithms are more amenable to vectorization and lead to similar Cray performance as noted for the infinite-depth case.

7.6   Time domain – infinite depth

For a fluid of infinite depth the potential of an impulsive source with strength $\delta(t)$ is

$$G = \frac{\delta(t)}{r} - \frac{\delta(t)}{r_1} + 2 \int_0^\infty \sqrt{gk} \sin(t\sqrt{gk})e^{kz} J_0(kR)dk \qquad (7.15)$$

(Wehausen & Laitone, 1960, eq. 13.49). Here the coordinates are as defined in §7.4.

It is convenient to use spherical coordinates (r_1, θ), with the angle θ measured from the negative vertical axis, and to nondimensionalize the physical parameters with respect to g and r_1. The variable t is replaced by $\tau = t\sqrt{g/r_1}$. Thus

$$G = \frac{\delta(t)}{r} - \frac{\delta(t)}{r_1} + g^{\frac{1}{2}} r_1^{-\frac{3}{2}} \operatorname{Re}\{F(\tau, \theta)\} \qquad (7.16)$$

where

$$F = -2i \int_0^\infty k^{\frac{1}{2}} \exp(ik^{\frac{1}{2}}\tau - k\cos\theta) J_0(k\sin\theta)dk$$

$$= -4i \int_0^\infty \omega^2 \exp(i\omega\tau - \omega^2\cos\theta) J_0(\omega^2\sin\theta)d\omega \qquad (7.17)$$

and the new integration variable is $\omega = k^{1/2}$. The objective is to evaluate (7.17) in the computational domain $(0 < \tau < \infty, \ 0 \leq \theta \leq \pi/2)$. The function F is the first-derivative with respect to time of the corresponding integral studied by Newman (1985b).

The integrand of (7.17) may be expanded in ascending powers of τ and integrated term-by-term to give the spherical-harmonic expansion

$$F = -2i \sum_{n=0}^{\infty} \frac{(i\tau)^n}{n!} \Gamma(\tfrac{1}{2}n + \tfrac{3}{2}) P_{\frac{1}{2}n+\frac{1}{2}}(\cos\theta) \tag{7.18}$$

where P_ν is the Legendre function. This expansion is uniformly convergent.

For large values of τ, asymptotic expansions can be derived in a similar but more complete manner relative to those outlined by Newman (1985b). For this purpose (7.17) is decomposed in the form

$$F = f_0 + f_1 + f_2 \tag{7.19}$$

where

$$f_0 = -4i \int_0^{\frac{1}{2}i\tau} \omega^2 \exp(i\omega\tau - \omega^2 \cos\theta) J_0(\omega^2 \sin\theta) d\omega \tag{7.20}$$

$$f_{1,2} = -4i \int_{\frac{1}{2}i\tau}^{\infty} \omega^2 \exp(i\omega\tau - \omega^2 \cos\theta) H_0^{(1,2)}(\omega^2 \sin\theta) d\omega \tag{7.21}$$

and $H_0^{(1,2)}$ denotes the Hankel functions $J_0 \pm iY_0$.

The integral f_0 can be expanded using Watson's Lemma, with the result

$$f_0 \simeq -4 \sum_{n=0}^{\infty} \frac{(2n+2)!}{n!} \tau^{-2n-3} P_n(\cos\theta) \tag{7.22}$$

The error in (7.22) is of order $\exp(-\tfrac{1}{4}\tau^2)$.

Let θ_0 denote a fixed angle such that $0 < \theta_0 < \tfrac{1}{2}\pi$. For $\theta < \theta_0$ the remaining integrals $f_{(1,2)}$ are exponentially small, and (7.22) is a complete asymptotic expansion of $F(\tau,\theta)$ for large τ.

For $\theta > \theta_0$ the argument of the Hankel functions in (7.21) is $O(\tau)$ or larger, provided the contour of integration is restricted so that $|\omega| \geq \tfrac{1}{2}\tau$. Thus these functions may be evaluated from the asymptotic expansions

$$H_0^{(1,2)}(\omega^2 \sin\theta) \simeq e^{\pm i(\omega^2 \sin\theta - \pi/4)} \sqrt{\frac{2}{\pi\omega^2 \sin\theta}}$$
$$\times \sum_{n=0}^{\infty} c_n \left(\frac{\pm i}{\omega^2 \sin\theta}\right)^n \tag{7.23}$$

where

$$c_n = \frac{[\Gamma(n+\tfrac{1}{2})]^2}{\pi 2^n n!} \tag{7.24}$$

Substituting (7.23) in (7.21) gives the expansions

$$f_{(1,2)} \simeq \frac{-4i}{\sqrt{2\pi \sin\theta}} e^{\mp i\pi/4} \sum_{n=0}^{\infty} (\mp i)^n c_n$$

$$\times \int_{\frac{1}{2}i\tau}^{\infty} \frac{\omega e^{(i\omega\tau \pm i\omega^2 \sin\theta - \omega^2 \cos\theta)}}{(\omega^2 \sin\theta)^n} d\omega \qquad (7.25)$$

The contours of integration in these two integrals may be deformed, in the half-plane $\mathrm{Re}(\omega) > 0$, with the upper limit a point at infinity in the sector $(-\pi/4 \pm \theta/2, \pi/4 \pm \theta/2)$. In each integral there is one saddle point, at $\omega_{(1,2)} = \frac{1}{2}\tau i e^{\pm i\theta}$. For the integral f_1 this saddle point is in the second quadrant and can be ignored. It follows that f_1 is exponentially small for all values of θ. For the integral f_2 the saddle point is in the first quadrant, and it is appropriate to deform the contour of integration to pass through this point. To clarify the analysis a local integration variable $v = ie^{-i\theta/2}(\omega/\omega_2 - 1)$ can be used to give

$$f_2 \simeq \frac{-4}{\sqrt{2\pi \sin\theta}} \exp(-\tfrac{1}{4}\tau^2 e^{-i\theta} + i\theta/2 + i\pi/4)$$

$$\times \sum_{n=0}^{\infty} c_n \frac{i^n \omega_2^{2-2n}}{(\sin\theta)^n} \int_{-2\sin(\theta/2)}^{\infty} e^{-\frac{1}{4}\tau^2 v^2} (1 - ie^{i\theta/2}v)^{1-2n} dv \quad (7.26)$$

After expanding the last factor in the integrand about $v = 0$, replacing the lower limit by $-\infty$, and integrating term-by-term, the desired asymptotic expansion is obtained in the form

$$f_2 \simeq \frac{-4i}{\sqrt{2 \sin\theta}} \exp(-\tfrac{1}{4}\tau^2 e^{-i\theta} - i\theta/2 + i\pi/4)$$

$$\times \sum_{n=0}^{\infty} \left(\frac{i}{\sin\theta}\right)^n \sum_{m=0}^{n} d_{nm} \omega_2^{1-2m-2n} e^{-im\theta} \qquad (7.27)$$

where $d_{00} = 1$, $d_{0m} = 0$ for $m > 0$, and for $n \geq 1$

$$d_{nm} = c_n \frac{(2m + 2n - 2)!}{(2n - 2)! 2^{2m} m!} \qquad (7.28)$$

If the substitution $\omega_2 = \frac{1}{2}i\tau e^{-i\theta}$ is made in (7.27) the result is a descending series in odd powers of the (large) parameter τ. Since the terms with $m + n \equiv N - 1$ are of order τ^{3-2N}, this double series may be summed in a triangular manner to any desired order N.

When $\theta = \pi/2$, F can be expressed in terms of products of Bessel functions of argument $\tau^2/8$ and orders $\pm\frac{1}{4}$ and $\pm\frac{3}{4}$, following the corresponding analysis of the Cauchy-Poisson problem (Wehausen & Laitone, 1960; see also Magee & Beck, 1989). For large values of τ these products can be expressed, by the trigonometric addition theorems, as the sum of a slowly-varying descending asymptotic series plus an oscillatory component, analogous respectively to (7.22) and (7.27).

Our first subroutine for the evaluation of this Green function used the three analytic expansions (7.18), (7.22), (7.27). The principal problem in this approach is that the partition between the ascending series (7.18) and asymptotic expansions (7.22), (7.27) must be at a relatively large value of τ, about $\tau = 9$ to obtain 6 decimals accuracy in the asymptotic expansions, but this is too large for efficient use of (7.18).

In our next subroutine we subdivided the (τ, θ) domain into 27 rectangles and approximated the difference between F and the asymptotic expansion (7.27) of 'optimum order' N in each subdomain by a two-dimensional economized polynomial. The optimum order N is defined to minimize the number of nonzero coefficients in the economized polynomial. Six decimals accuracy is achieved with an average of 30 nonzero polynomial coefficients in each subdomain. This approach, with the number of terms in the asymptotic expansion 'optimized' between 0 and 6, appeared to be an elegant combination of analytical and numerical techniques, and it led to a significant improvement in terms of computational speed. However the $N = 6$ steepest-descent evaluation (involving a total of 21 complex terms) required substantial time.

The third subroutine developed for this Green function reflects the experience with the more 'heavy-handed' use of economized polynomials in the frequency domain. The objective was to restrict the order of the steepest-descent expansion to $N = 2$, without sacrificing accuracy. This has been achieved by using a finer rectangular grid with a total of 240 subdomains, defined with partitions at $\tau = 2(.5)13$ and $\cos \theta = 0.1(.1)0.9$. In each subdomain the economized polynomial requires 32 coefficients, hence a total of 7680 coefficients must be stored. The polynomial is augmented by the steepest-descent expansion with $N = 2$ in the subdomains where $\tau > 4$ and $\cos \theta < .4$. The coding is similar to that described in the frequency domain, and vectorization is feasible on the Cray. (For the Green function and its two first-derivatives, the time required on a VAX 11-750 is about 1 millisecond; on the Cray with vectorization implemented the corresponding time is about 1 microsecond. These times are somewhat faster than we have achieved for the frequency-domain Green function.)

In deriving the Chebyshev expansion coefficients, F is evaluated by a combination of four separate algorithms: (a) for small or moderate values of τ (7.18) is used; (b) for very large values of τ the asymptotic expansions (7.22) and (7.25) are used; (c) for intermediate values of τ, and $\cos \theta > \frac{1}{2}$, (7.17) is integrated numerically along the real axis; and (d) for intermediate values of τ and smaller values of $\cos \theta$ the integral is evaluated numerically in the complex plane. The numerical integration is performed by Romberg quadratures with a convergence tolerance of 10^{-10}.

7.7 Time domain – finite depth

In a fluid of depth h, the impulsive Green function is represented in the form

$$\begin{aligned}
G &= \frac{\delta(t)}{r} + \frac{\delta(t)}{r_2} \\
&\quad - 2\delta(t) \int_0^\infty \frac{e^{-kh}}{\cosh(kh)} \mathcal{G}(y,z,R)\,dk \\
&\quad + 2 \int_0^\infty \frac{\sqrt{gk\tanh kh}}{\cosh(kh)\sinh(kh)} \sin(t\sqrt{gk\tanh kh})\mathcal{G}(y,z,R)\,dk,
\end{aligned}$$
(7.29)

(Wehausen & Laitone, 1960, eq. 13.53).

It is convenient to nondimensionalize with respect to the parameters g and h. After defining the variables $X = R/h$, $Y = -\zeta/h$, $Z = -z/h$, $T = t\sqrt{g/h}$, and using the addition theorem for the product of two hyperbolic cosines, G can be expressed in the form

$$\begin{aligned}
G &= g^{1/2}h^{-3/2}\delta(T)[F_0(X,Y-Z) + F_0(X,2-Y-Z)] \\
&\quad + g^{1/2}h^{-3/2}[F(X,Y-Z,T) + F(X,2-Y-Z,T)]
\end{aligned}$$
(7.30)

where

$$F_0(X,V) = \frac{1}{\sqrt{X^2+V^2}} - \int_0^\infty e^{-k}\operatorname{sech}k \cosh(kV)J_0(kX)\,dk$$
(7.31)

and

$$\begin{aligned}
F(X,V,T) &= \int_0^\infty \frac{\sqrt{k\tanh k}}{\cosh k \sinh k} \sin(T\sqrt{k\tanh k}) \\
&\quad \times \cosh(kV)J_0(kX)\,dk
\end{aligned}$$
(7.32)

The vertical coordinate V is in the range $(0,2)$. It is straightforward to evaluate $F_0 = G_-(X,V) - \log(2)$ using (7.4), (7.8), and the polynomial coefficients in Table 7.3.

The efficient evaluation of the time-dependent function F presents a more challenging problem. As in the frequency domain it is useful in some regimes to subtract from (7.32) the 'infinite-depth limit' F_∞, defined by replacing the hyperbolic functions in (7.32) by their respective limiting forms for large k. This limit can be evaluated by means of the algorithms described in §7.6.

For values of the nondimensional time T up to about 10 it is possible to approximate either the function F or the difference $F - F_\infty$ using economized triple polynomials in the variables X, V, T. Preliminary results have been obtained for the relevant range $(0 \le V \le 2)$ with subdomains in the X,T space bounded by partitions at integer intervals. For example, for $(6 \le X \le 7)$ and $(7 \le T \le 8)$ F can be approximated with about 6 in the X, T space bounded by partitions at integer intervals. For example, for $(6 \leq X \leq 7)$ and $(7 \leq T \leq 8)$ F can be approximated with about 6 decimals accuracy by a polynomial with 125 terms. In the same domain the difference $F - F_\infty$ can be approximated with 118 terms. The latter function is the appropriate quantity to use for smaller values of X. In applications where the solution proceeds in time steps, the nondimensional time T is constant at each step and for each combination of source and field points a relatively simple polynomial is evaluated. Thus in the last example with 118 terms in the triple polynomial, pre-evaluation with a fixed value of T yields a polynomial in (X, V) with only 31 terms. (This procedure is similar to that described in §7.5, where T is replaced by Kh.)

Asymptotic expansions are required for larger values of time, to provide guidance in the development of numerical algorithms. As in the analogous treatment of the two-dimensional Cauchy-Poisson problem by Whitham (1974), it is necessary to consider three overlapping domains in relation to the front $X = T$.

Well behind the front, where $X << T$, (7.32) can be expressed in a form analogous to (7.19)-(7.21). Thus

$$F(X, V, T) = \mathrm{Re}\{f_0 + f_1 + f_2\} \tag{7.33}$$

where

$$f_0 = -2i \int_0^{i\kappa_0} \frac{\omega}{\sinh 2k} \cosh kV e^{i\omega T} J_0(kX) dk \tag{7.34}$$

$$f_{1,2} = -i \int_{i\kappa_0}^{\infty} \frac{\omega}{\sinh 2k} \cosh kV e^{i\omega T} H_0^{(1,2)}(kX) dk \tag{7.35}$$

and $\omega(k) = \sqrt{k \tanh k}$. The parameter κ_0 may be fixed so that the upper limit in (7.34) is at an arbitrary point below the first pole on the positive imaginary axis, say $\kappa_0 = \pi/4$.

As in §7.6, these three integrals contribute, respectively, a slowly-varying asymptotic series, an exponentially small contribution, and a contribution from the saddle point on the real axis.

We begin by considering the function f_0 defined by (7.34). Since $\omega(k) \simeq k$ is single-valued near the origin it is straightforward to substitute $k = i\kappa$ and $\Omega(\kappa) = -i\omega(k) = \sqrt{\kappa \tan \kappa}$. It follows that

$$f_0 = 2 \int_0^{\kappa_0} \frac{\Omega}{\sin 2\kappa} \cos \kappa V e^{-\Omega T} I_0(\kappa X) d\kappa \tag{7.36}$$

where I_0 is the modified Bessel function of the first kind. Since $T > X$ and $\Omega \geq \kappa$ the integrand of (7.36) is exponentially small away from the origin. To estimate the contribution near the origin we use the series expansion

$$\Omega = \kappa + \tfrac{1}{6}\kappa^3 + O(\kappa^5) \tag{7.37}$$

and expand the integrand of (7.36) in the form

$$f_0 = \int_0^{\kappa_0} \left[1 + \left(\tfrac{5}{6} - \tfrac{1}{2}V^2 \right) \kappa^2 - \tfrac{1}{6}T\kappa^3 + O(\kappa^4) \right] e^{-\kappa T} I_0(\kappa X) d\kappa \tag{7.38}$$

After replacing the upper limit by ∞ and using the transform

$$\int_0^\infty e^{-\kappa T} I_0(\kappa X) d\kappa = (T^2 - X^2)^{-\frac{1}{2}} \tag{7.39}$$

(and its partial derivatives with respect to T) to evaluate (7.38) term-by-term, it follows that

$$\begin{aligned} f_0 \;=\; & (T^2 - X^2)^{-\frac{1}{2}} \\ + \; & \left(\tfrac{5}{6} - \tfrac{1}{2}V^2 \right) (2T^2 + X^2)(T^2 - X^2)^{-\frac{5}{2}} \\ - \; & \tfrac{1}{2}T^2(2T^2 + 3X^2)(T^2 - X^2)^{-\frac{7}{2}} \\ + \; & O(T^4[T^2 - X^2]^{-\frac{9}{2}}) \end{aligned} \tag{7.40}$$

This sequence converges asymptotically, subject to the inequalities $(T - X)^3 \gg T \gg 1$.

Proceeding to the asymptotics of (7.35), we note that in the strip $-\tfrac{1}{2}\pi < \text{Im}(k) < \tfrac{1}{2}\pi$ adjacent to the positive real axis, the imaginary part of $w(k)$ is bounded and has the same sign as $\text{Im}(k)$. From (7.23) it follows that f_1 can be evaluated along the contour $\text{Im}(k) = \kappa_0$, where the integrand is exponentially small and thus f_1 is of order $e^{-\kappa_0 T}$. For f_2 there is a saddle point on the real axis at the (positive) root of the equation $w' = X/T$, where the integrand is locally of order one. At this point the path of steepest descent is along the vector which intersects the real k-axis at an angle $-\pi/4$. Defining the root by k_0 and the corresponding value $w_0 = w(k_0)$, a procedure similar to that in §7.6 gives the leading-order approximation

$$\text{Re}(f_2) \simeq 2\sqrt{\frac{\tanh k_0}{|w_0''|XT} \frac{\cosh k_0 V}{\sinh 2k_0}} \sin(w_0 T - k_0 X) \tag{7.41}$$

where w_0'' is the (negative) second derivative of $w(k)$ evaluated at $k = k_0$. An algorithm for extending (7.41) to higher-order terms can be constructed in a manner analogous to (7.27), but the analysis is involved. To illustrate this procedure the two leading terms after (7.41) are derived in the Appendix.

Proceeding in a similar manner ahead of the front, for $X \gg T$, the saddle point is on the imaginary axis at $k_0 = i\kappa_0$ defined such that $\Omega(\kappa) = \sqrt{\kappa \tan \kappa}$ and $\Omega' = X/T$. Note that $(0 < \kappa_0 < \pi/2)$. After writing (7.32) in the form

$$F(X, V) = \text{Im} \int_0^\infty \frac{w}{\sinh 2k} \cosh kV \left(e^{iwt} - e^{-iwt} \right) H_0^{(1)}(kX) dk \tag{7.42}$$

the contour of integration may be deformed to pass up the imaginary axis, from 0 to $i\kappa_0$, and then parallel to the real axis to $k = \infty + i\kappa_0$. The integral along the imaginary axis is real and does not contribute to (7.42). For the integral parallel to the real axis the asymptotic expansion (7.23) may be used for the Hankel function, confirming that the lower limit of this integral is a saddle point and the path of steepest descent is orthogonal to the imaginary axis. It follows that

$$F(X,V) \simeq \sqrt{\frac{\tan\kappa_0}{\Omega_0'' XT}} \frac{\cos\kappa_0 V}{\sin 2\kappa_0} \exp(\Omega_0 T - \kappa_0 X) \tag{7.43}$$

The difference between the oscillatory and exponential behaviour is obvious in (7.41) and (7.43). It should be noted that the slowly-varying component (7.36) is present only behind the front. Higher-order terms in (35) can be derived in a manner analogous to (7.41).

It remains to consider the transition region near the front, where $X \simeq T \gg 1$. In this case the saddle point approaches the origin $k = 0$, and asymptotic expansion of the Bessel function is not justified. In the vicinity of $k = 0$ the dispersion relation is approximated by

$$\omega = k - \frac{k^3}{6} + O(k^5) \tag{7.44}$$

Using the new integration variable $u = \sinh^{-1}(k)$, and expanding the integrand about $u = 0$, it follows that

$$F = \int_0^{} [1 - \tfrac{1}{3}u^2 + \tfrac{1}{2}u^2 V^2 + O(u^4)]$$
$$\times \sin(uT + O(u^5 T)) J_0(X \sinh u) du \tag{7.45}$$

The leading-order contribution

$$F \simeq \int_0^\infty \sin(uT) J_0(X \sinh u) du$$
$$= \frac{1}{\pi} \sinh(\tfrac{\pi}{2}T) \left[K_{\frac{1}{2}iT}(\tfrac{1}{2}X)\right]^2 \tag{7.46}$$

is evaluated from a variant of Nicholson's integral (Erdélyi, et al, 1954, §2.13, eq. 58), where K denotes the modified Hankel function of imaginary order $\tfrac{1}{2}iT$. The next term in the asymptotic expansion of F, associated with the $O(u^2)$ terms in (7.45), can be evaluated from the second derivative of (7.46) with respect to T. This procedure may be extended to include terms of higher order.

A uniform asymptotic expansion of the modified Hankel function in (7.46) has been derived in the transition region by Balogh (1967). In the form given by Olver (1974, p. 425) the leading term is

$$K_{i\nu}(\nu z) \simeq \pi\nu^{-\frac{1}{3}} e^{-\nu\pi/2} \left(\frac{4\zeta}{1-z^2}\right)^{\frac{1}{4}} \mathrm{Ai}(-\nu^{\frac{2}{3}}\zeta) \tag{7.47}$$

where Ai is the Airy integral and the parameter ζ is defined by

$$\tfrac{2}{3}\zeta^{\frac{3}{2}} = \log\left(\frac{1+\sqrt{1-z^2}}{z}\right) - \sqrt{1-z^2}$$

$$= \tfrac{1}{3}(1-z^2)^{\frac{3}{2}} + O(1-z^2)^{\frac{5}{2}} \tag{7.48}$$

In the present case $\nu = \tfrac{1}{2}T$ and $z = X/T$. With these substitutions in (7.46) it follows that

$$F \simeq 2^{\frac{1}{3}}\pi T^{-\frac{2}{3}}\left[\mathrm{Ai}\left(-(\tfrac{1}{4}T)^{\frac{2}{3}}(1-X^2/T^2)\right)\right]^2 \tag{7.49}$$

If the asymptotic expansions of the Airy integral are used, for large negative or positive values of its argument, (7.49) can be shown to overlap with the corresponding approximations of (7.40) plus (7.41), or (7.43), respectively, for small values of k_0 or κ_0. Higher-order terms can be obtained by use of the complete asymptotic expansion of the modified Hankel function (Olver, 1974, p. 425), and the procedure outlined following (7.46).

It is interesting to note that the square of an Airy function is involved here, giving rise to both oscillatory and slowly-varying components in a similar manner to the infinite-depth limit discussed in §7.6. This feature of the far-field wave motion differs from the analogous two-dimensional case, where the Airy function enters in a linear manner and the slowly-varying component is absent (Whitham, 1974).

Some computational results are shown in Figures 7.1-7.2 and Table 7.7 for the function $\Phi(X,T) = F(X,2,T) + F(X,0,T)$, equivalent to the nondimensionalized source potential (7.29) when the source and field points are both in the free surface and $t > 0$. The computations are based on numerical integration of the difference between (7.32) and the infinite-depth integral F_∞. Figure 7.1 shows a sequence of 'snapshots' for $T = 1, 2, 3, ..., 16$. The advance of the front with unit velocity is apparent, as is the absence of waves ahead of the front. The short-wavelength oscillations for small values of X are an obvious feature. Figure 7.2 shows the difference $\Phi - \Phi_\infty$, where the short waves are effectively removed.

On a historical note, Rayleigh (1909) considered the solution of the two-dimensional Cauchy-Poisson problem in both finite and infinite depth, and contrasted the far-field asymptotic behaviour of the two cases. The closing paragraph in that work is noteworthy:

> If we attempt to fill up the gaps in our solution by applying quadratures ... we have to face the difficulty that, as written, the integral is not convergent. Some analytical transformation is called for. One way out of the difficulty might be to calculate the difference between the solutions for finite and infinite depth, for it would appear that the nonconvergent part of the integral, corresponding to infinitely small wavelengths, must be the same for both.

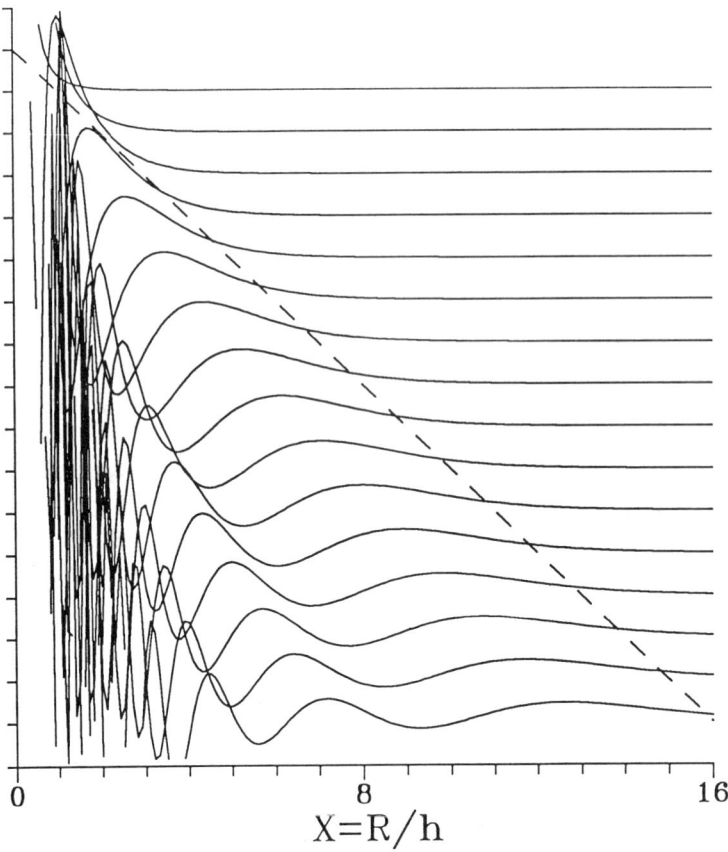

Figure 7.1: Sequence of values of the source potential (7.29) for nondimensional times $T = t\sqrt{g/h} = 1, 2, \ldots, 16$. Each successive curve is displaced downard by a unit increment to indicate the progressive advance of the waves. The front $X = T$ is indicated by the dashed line.

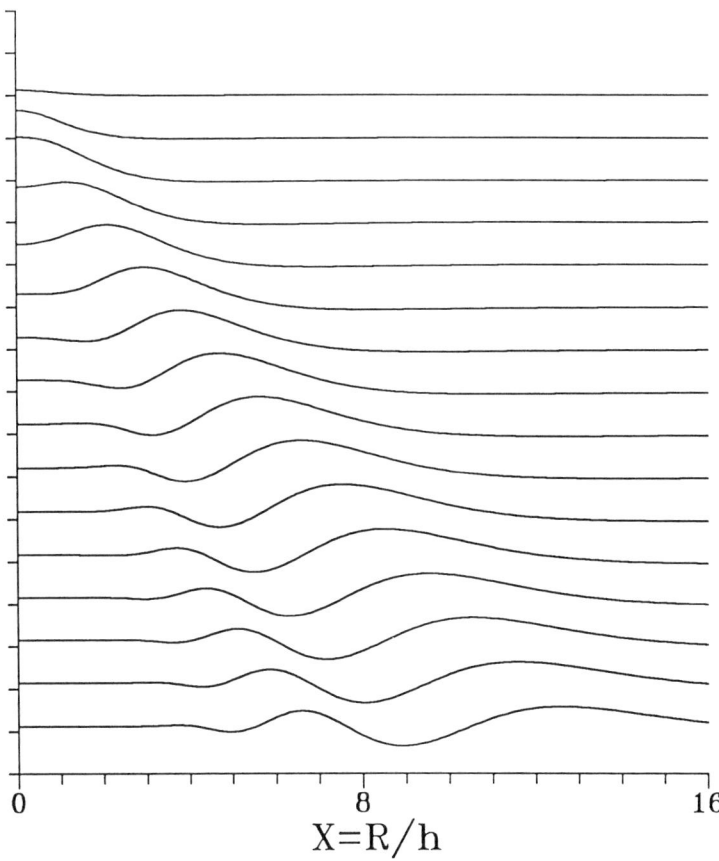

$$X=R/h$$

Figure 7.2: Sequence of values of the difference integral (finite-minus infinite-depth source potential) for nondimensional times $T = t\sqrt{g/h} = 1, 2, \ldots, 16$. Each successive curve is displaced downard by a unit increment to indicate the progressive advance of the waves.

Table 7.7 lists the computational results for $T = 10$, as well as the corresponding asymptotic approximations derived from the leading term of (7.40) plus (7.41) for points behind the front, from (7.43) ahead of the front, and from (7.49) near the front. From comparison of the last three columns it is apparent that the asymptotic approximations are qualitatively correct, but more complete asymptotic expansions are required for numerical purposes.

7.8   Discussion and conclusions

Several examples have been given to show the utility of multivariate polynomial approximations in the evaluation of free-surface Green functions. The algorithms described have lead to substantial reductions in computing time, particularly on vector computers where the systematic use of polynomials with the same form in different subdomains of the

X	$\Phi - \Phi_\infty$	Φ_∞	Φ	Φ_A	(7.49)
1	0.208058	-1.809516	-1.601458	-1.771233	0.184112
2	0.255121	-0.206861	0.048260	-0.132532	0.216880
3	0.135583	1.375727	1.511310	1.413654	0.271550
4	-0.110938	-0.018690	-0.129628	-0.199518	0.343613
5	0.301172	-0.545149	-0.243976	-0.369075	0.419613
6	0.769164	-0.348449	0.420715	0.282065	0.476037
7	0.809345	-0.135816	0.673529	0.563328	0.486309
8	0.576039	-0.011280	0.564759	0.506230	0.435689
9	0.311478	0.048964	0.360443	0.386268	0.334335
10	0.122172	0.073451	0.195623		0.214972
11	0.015176	0.079871	0.095048	0.147248	0.113402
12	-0.035201	0.077758	0.042557	0.055907	0.048079
13	-0.054106	0.071988	0.017882	0.021917	0.016053
14	-0.057846	0.064986	0.007139	0.008422	0.004137
15	-0.055178	0.057910	0.002732	0.003146	0.000806
16	-0.050274	0.051283	0.001009	0.001142	0.000117
17	-0.044942	0.045303	0.000361	0.000404	0.000012
18	-0.039888	0.040014	0.000126	0.000140	0.000001
19	-0.035342	0.035385	0.000043	0.000047	0.000000
20	-0.031340	0.031354	0.000014	0.000016	0.000000

Table 7.4: Values of the difference integral, infinite-depth component, and total finite-depth potential $\Phi(X, T) = F(X, 2, T) + F(X, 0, T)$ for $T = 10$. The source and field points are both in the free surface. Φ_A denotes the asymptotic approximation evaluated behind the front $X = T$ from the leading term in (7.40) plus (7.41), and ahead of the front from (7.43). The last column, evaluated from (7.49), is the asymptotic approximation valid near the front.

computational space permits the simultaneous evaluation of a large array of Green functions within a single vectorized loop.

The use of economized polynomials is effective for evaluating those components of the Green functions which are slowly-varying. Analysis is required to develop suitable forms for representing singular features, as in the simpler case of a typical special function such as the Bessel functions. Appropriate forms for this purpose have been indicated in §§7.4-7.6. For the finite-depth transient Green function, the large-time asymptotic analysis in §7.7 is expected to serve as the basis for efficient numerical approximations, but that extension is left for future work.

An alternative to the use of polynomial approximations is to interpolate in a precomputed table. The later technique has been utilized for the infinite-depth Green functions by Delhommeau (1989) in the frequency domain, and Ferrant (1988) and Magee & Beck (1989) in the time domain. Large tables are required in these works, ranging from 64,000 to 2,000,000 entries. Thus the memory requirements are greater, and substantial time may be involved in table look-up. Comparisons of computing time are difficult to resolve due to differences in the accuracy achieved, and in the computing systems used.

There is of course a fundamental similarity between interpolation in a table and the alternative use of economized polynomials. In the most extreme case, where the number of subdomains is sufficiently large to permit the use of economized polynomials of degree one, the result is practically equivalent to linear interpolation in a table. Similarly, if some quadratic terms are included, the resulting algorithm is closely related to the bicubic interpolation scheme of Magee & Beck (1989). In both cases there is a minor advantage for the economized polynomial in that it approximates to a minimax fit, but that feature is not important at such a low degree. The most important advantages of the economized polynomial appear to be (a) that it permits a wide option of choices compromising between subdivision of the computational domain and the degree of the polynomials; and (b) that when the polynomial degree is substantial the minimax approximation increases the computational efficiency.

In the examples described here polynomials of maximum degree 5-10 have been used, with $O(100)$ subdivisions of the computational domain. These choices have been based at least in part on the desire to restrict the total number of polynomial coefficients to a few thousand, and we have not systematically studied the effect on computing efficiency of substantially larger numbers of subdivisions with simpler polynomials in each one. Such a comparison obviously would depend on the particular computing system utilized, more specifically on the relative speed of floating-point operations *vs.* random access to memory.

The free-surface Green functions are more difficult to evaluate when a uniform horizontal translation is imposed to account for the forward velocity of a ship. Since the flow is not axisymmetric, an additional coordinate is required in approximations or tables. A more fundamental difficulty is the existence of an essential singularity downstream in the plane of the free surface, when the source point is in the same plane. Finally, it is difficult to derive asymptotic expansions in the far field with uniform accuracy, particularly near the outer limits of the Kelvin wave system. The Green function consists of two distinct components, a double integral which is symmetric about the axis of translation and evanescent, and a single integral which describes the wavelike features downstream. The double integral appears superficially to be more difficult to evaluate, but in fact the converse is the case. Effective triple polynomials for evaluating the double integral have been tabulated by Newman (1987). For the single integral Neumann expansions originally derived by Bessho are effective in part of the computational domain (Baar & Price, 1988), but not far downstream or close to the essential singularity on the axis. These difficult domains are discussed by Ursell (1960, 1988), Newman (1988) and Clarisse (1989, 1990). The generalization where oscillatory time-dependence is superposed with uniform translation is considered by Ohkusu & Iwashita (1989), using a numerical implementation of the method of steepest descent.

The mathematical difficulties which result from a steady forward velocity can be circumvented by analysing the moving steady-state source as a sequence of impulsive Green functions, in essentially the same manner that Kelvin originally analysed ship waves. The Green functions considered in §§7.6-7.7 are applicable, provided the computations can be performed over a sufficient time to reach a steady state. Numerical examples for a point source are given by Newman (1985b), and substantial progress using this approach has been made by King *et al* (1988).

In two distinct areas free-surface Green functions have been abandoned in favour of distributions of Rankine singularities on the free surface. The first is the numerical solution of fully-nonlinear wave/body problems in the time domain, where impressive progress has been made by several workers in the past decade. However these solutions are for idealized bodies, primarily in two dimensions, and applications for general bodies in three dimensions appear at present to be impractical. The second area is the analysis of three-dimensional steady-state ship waves and wave resistance, based on the method originally developed by Dawson. Here practical results have been achieved, and recent efforts have been made to extend this approach to oscillatory motions in waves. There are fundamental limitations on the ability of this method to compute waves in the far field, and to solve the oscillatory problem in the low-frequency regime where waves are radiated upstream. In both areas there are potential advantages in hybrid solutions, combining Rankine distributions in the near field with free-surface Green functions on an exterior matching boundary.

The numerical analysis of free-surface Green functions is challenging from the complementary viewpoints of classical analysis, asymptotics, and approximation theory. The results obtained in recent years have facilitated the hydrodynamic analysis of offshore platforms. More work remains, particularly in the case where uniform translation is involved, but the possibility of robust panel methods for analysing moving ships seems much closer to a reality than was true a few years ago.

7.9 Acknowledgement

Portions of this work have been supported by the Office of Naval Research, the Exxon Production Research Company, and by a Joint Industry Project which includes sponsorship from Det norske Veritas, Norsk Hydro AS, Shell Development Company and Statoil AS.

7.10 References

Abramowitz, M., & Stegun, I. A., 1964 *Handbook of Mathematical Functions with Formulas, Graphs, and Mathematical Tables*, Government Printing Office, Washington and Dover, New York.

Baar, J. J. M. & Price, W. G., 1988 Evaluation of the wavelike disturbance in the Kelvin wave source potential, *J. Ship Research* **32**, 1, 44-53.

Balogh, C. B., 1967 Asymptotic expansions of the modified Bessel function of the third kind of imaginary order. *SIAM J. Appl. Math.* **15**, 1315-1323.

Breit, S. R., 1990 The potential of a Rankine source between parallel planes and in a rectangular cylinder, *Journal of Engineering Mathematics*, to appear.

Chester, C., Friedman, B., & Ursell, F., 1957 An extension of the method of steepest descents, *Proc. Camb. Phil. Soc.* **53**, 599-611.

Clarisse, J. M., 1989 On the numerical evaluation of the Neumann-Kelvin Green function, M.S. Thesis, MIT, Cambridge, MA.

Clarisse, J. M., 1990 Thirty years after... the evaluation of the single integral part of the Kelvin wave source potential in the far-field. Fifth International Workshop on Water Waves and Floating Bodies, Manchester.

Delhommeau, G., 1989 Amélioration des performances des codes de calcul de diffraction-radiation au premier ordre. 2èmes Journées de l'Hydrodynamique, Nantes.

Erdélyi, A., Magnus, W., Oberhettinger, F., & Tricomi, F. G., 1954 *Tables of integral transforms*, **1**, McGraw-Hill, New York.

Ferrant, P. 1988 Radiation d'ondes de gravité par les mouvements de grande amplitude d'un corps immergé. Thèse de Doctorat, Université de Nantes.

Gradshteyn, I. S. & Ryzhik, I. M. 1965 *Tables of Integrals, Series and Products*, Academic Press, New York.

Havelock, T. H. 1955 Waves due to a floating sphere making periodic heaving oscillations, *Proc. Roy. Soc.* A, **231**, 1-7.

132   *J.N. Newman*

King, B. K., Beck, R. F. & Magee, A. R. 1988 Seakeeping calculations with forward speed using time-domain analysis. 17th Symposium on Naval Hydrodynamics, The Hague.

Magee, A. R. & Beck, R. F. 1989 Vectorized computation of the time-domain Green function. Fourth International Workshop on Water Waves and Floating Bodies, Øystese, Norway.

Newman, J. N. 1985a Algorithms for the free-surface Green function, *Journal of Engineering Mathematics*, **19**, 57-67.

Newman, J. N. 1985b The evaluation of free-surface Green functions, Fourth International Conference on Numerical Ship Hydrodynamics, Washington.

Newman, J. N. 1987 Evaluation of the wave-resistance Green function - Part 1 - The double integral, *Journal of Ship Research*, **31**, 2.

Newman, J. N. 1988 Evaluation of the wave-resistance Green function near the singular axis, Third International Workshop on Water Waves and Floating Bodies, Woods Hole.

Ohkusu, M. & Iwashita, H. 1989 Evaluation of the Green function for ship motions at forward speed and application to radiation and diffraction problems, Fourth International Workshop on Water Waves and Floating Bodies, Øystese, Norway.

Olver, F. W. J., 1974 *Asymptotics and special functions*, Academic Press, New York.

Rayleigh, Lord 1909 On the instantaneous propagation of disturbance in a dispersive medium, exemplified by waves on water deep and shallow, *Phil. Mag.*, Series 6, **18**, 1-6.

Ursell, F. 1949 On the heaving motion of a circular cylinder on the surface of a fluid, *Quart. J. Mech. Appl. Math.* **2**, 218-31.

Ursell, F. 1960 On Kelvin's ship-wave pattern, *J. Fluid Mech.* **8**, 418-31.

Ursell, F. 1988 On the Kelvin wave-source potential, Third International Workshop on Water Waves and Floating Bodies, Woods Hole.

Wehausen, J. V. & Laitone, E. V. 1960 Surface waves, *Handbuch der Physik*, **9**, 446-778. Springer-Verlag, Berlin.

Whitham, G. B. 1974 *Linear and Nonlinear Waves*, Wiley-Interscience, New York.

7.11   Appendix – extension of equation 7.41

An algorithm for extending (7.41) to higher-order terms can be constructed for $T \gg 1$ in a manner analogous to (7.26). It is convenient to define $\omega_n = [d^n/dk^n \omega(k)]_{k=k_0}$. Note that $\omega_1 \equiv \omega_0'$ is positive, decreasing monotonically from 1 to 0 as k increases along the positive real axis, and $\omega_2 \equiv \omega_0''$ is negative. For $X/T = O(1)$, $k_0 = O(1)$ whereas if $X/T \ll 1$, $k_0 \simeq (T/2X)^2$. Thus for all values of $0 < X < T$ the product $k_0 X \gg 1$, and the asymptotic expansion (7.23) may be used near the saddle point

After making this substitution,

$$f_2 \simeq -i\sqrt{\frac{2}{\pi X}}e^{i\pi/4}\sum_{n=0}(-i)^n c_n X^{-n}$$

$$\times \int_{i\kappa_0}^{\infty}\frac{\omega}{\sinh 2k}\cosh kV e^{i\omega T-ikX}(k)^{-n-\frac{1}{2}}\,dk$$

$$\equiv -i\sqrt{\frac{2}{\pi X}}e^{i\pi/4}\sum_{n=0}(-i)^n c_n X^{-n}\int_{i\kappa_0}^{\infty}\lambda_n(k)e^{i\omega T-ikX}\,dk \qquad (7.50)$$

To expand the last integral, we first define a local variable $u = e^{i\pi/4}(k-k_0)$, together with the Taylor series expansion

$$i\omega(k)T - ikX = i\omega_0 T - ik_0 X + iT\sum_{m=2}\frac{\omega_m}{m!}u^m e^{-im\pi/4} \qquad (7.51)$$

and define the new variable v such that

$$i\sum_{m=2}\frac{\omega_m}{m!}u^m e^{-im\pi/4} = -\frac{1}{2}|\omega_2|v^2 \qquad (7.52)$$

Thus

$$v = \sum_{m=0}a_m u^{m+1}e^{-im\pi/4} \qquad (7.53)$$

where $a_0 = 1$. Reverting the last series gives a relation for $u(v)$ in the form

$$u = \sum_{m=0}b_m v^{m+1}e^{-im\pi/4} \qquad (7.54)$$

where $b_0 = 1$. The first three coefficients in these series are defined as follows:

$$
\begin{aligned}
a_0 &= 1 \\
a_1 &= \frac{1}{6}(\omega_3/\omega_2) \\
a_2 &= \frac{1}{24}(\omega_4/\omega_2) - \frac{1}{72}(\omega_3/\omega_2)^2 \\
b_0 &= 1 \\
b_1 &= -a_1 = -\frac{1}{6}(\omega_3/\omega_2) \\
b_2 &= 2a_1^2 - a_2 = \frac{5}{72}(\omega_3/\omega_2)^2 - \frac{1}{24}(\omega_4/\omega_2) \qquad (7.55)
\end{aligned}
$$

The last integral in (7.50) can be expanded in the form

$$\int_{i\kappa_0}^{\infty}\lambda_n(k)e^{i\omega T-ikX}\,dk$$

$$\simeq \quad e^{i\omega_0 T - ik_0 X} \int_{-\infty}^{\infty} \frac{\lambda_n(k)}{dv/dk} \exp(-\frac{1}{2}|\omega_2|Tv^2)\, dv$$

$$= \quad e^{i\omega_0 T - ik_0 X - i\pi/4} \int_{-\infty}^{\infty} \sum_{m=0} d_n^m v^m e^{-im\pi/4} \exp(-\frac{1}{2}|\omega_2|Tv^2)\, dv$$

$$= \quad e^{i\omega_0 T - ik_0 X - i\pi/4} \sqrt{\frac{2\pi}{|\omega_2|T}} \sum_{m=0} d_n^{2m} \frac{(2m-1)!!}{(|\omega_2|T)^m} e^{-im\pi/4} \qquad (7.56)$$

where $(2m-1)!! = (2m-1)\cdot(2m-3)\cdot ... \cdot 3\cdot 1$, and the coefficients d_n^m are defined by the series

$$\lambda_n(k)\frac{du}{dv} = \sum_{m=0} d_n^m v^m e^{-im\pi/4} \qquad (7.57)$$

Substitution of (7.56) in (7.50) gives the desired asymptotic expansion

$$f_2 \quad \simeq \quad -2i e^{i\omega_0 T - ik_0 X} \sqrt{\frac{1}{|\omega_2|XT}}$$

$$\times \quad \sum_{n=0}(-i)^n c_n X^{-n} \sum_{m=0} d_n^{2m} \frac{(2m-1)!!}{(|\omega_2|T)^m} e^{-im\pi/4} \qquad (7.58)$$

In the limit where $X/T \to 0$, $k_0 \to \infty$ and (7.58) is equivalent to the infinite-depth expansion (7.23).

The term in (7.58) with $m = n = 0$ is equal to (7.41). Higher-order terms can be evaluated in principle, but the algebra is cumbersome for $m > 0$. The two terms $(m = 0, n = 1)$ and $(m = 1, n = 0)$, which are of comparable order, require the coefficients

$$d_1^0 = \lambda_1(k_0) = \frac{\omega_0}{\sinh 2k_0} \cosh(k_0 V)(k_0)^{-\frac{3}{2}} \qquad (7.59)$$

and

$$d_0^2 = \tfrac{1}{2}\lambda_0'' + 3b_1\lambda_0' + 3b_2\lambda_0 \qquad (7.60)$$

where primes denote differentiation with respect to the variable k_0. Expressions for evaluating these derivatives are given below.

$$\lambda_0 \quad = \quad \frac{\cosh(k_0 V)}{\sinh(2k_0)}\omega_0 k_0^{-\frac{1}{2}}$$

$$\lambda_0' \quad = \quad \frac{\cosh(k_0 V)}{\sinh(2k_0)}\left(\omega_1 k_0^{-\frac{1}{2}} - 2\coth(2k_0)\omega_0 k_0^{-\frac{1}{2}} - \tfrac{1}{2}\omega_0 k_0^{-\frac{3}{2}}\right)$$

$$\qquad + \quad \frac{\sinh(k_0 V)}{\sinh(2k_0)}\omega_0 V k_0^{-\frac{1}{2}}$$

$$\lambda_0'' \quad = \quad \frac{\cosh(k_0 V)}{\sinh(2k_0)}\left(-4\omega_0 k_0^{-\frac{1}{2}} + \omega_0 V^2 k_0^{-\frac{1}{2}} + \tfrac{3}{4}\omega_0 k_0^{-\frac{5}{2}}\right.$$

$$- \quad \omega_1 k_0^{-\frac{3}{2}} + \omega_2 k_0^{-\frac{1}{2}} + \coth(2k_0)(2\omega_0 k_0^{-\frac{3}{2}} - 4\omega_1 k_0^{-\frac{1}{2}})$$

$$+ \quad 8\coth^2(2k_0)\omega_0 k_0^{-\frac{1}{2}} \bigg)$$

$$+ \quad \frac{\sinh(k_0 V)}{\sinh(2k_0)} V \left(-\omega_0 k_0^{-\frac{3}{2}} + 2\omega_1 k_0^{-\frac{1}{2}} - 4\coth(2k_0)\omega_0 k_0^{-\frac{1}{2}} \right)$$

8

On far-field approximations to the wave pattern around a ship travelling at constant velocity

K. Eggers
University of Hamburg

... I am beginning to doubt whether the concept of a free wave in an infinitely wide canal will ultimately prove useful.
F. Ursell, Ann Arbor, 1963.

8.1   Historical remarks

As long as man has used naval transportation, he must have observed that moving floating bodies make waves, which reflect some part of the effort needed to keep a vessel advancing in the desired direction. But it was only 40 years ago that interest awakened in determining the power supplied to the wave system from the ship through evaluation of wave pattern characteristics. In general, the creation of wave patterns at model scale was just a game. Even when renowned mathematicians found interest in the subject of water waves, they just used it as a tool for applying and inventing analytical methods. This applies in particular to a fundamental paper of A.-L. Cauchy (1827) entitled 'Théorie de la propagation des ondes à la surface d'un fluide pesant d'une profondeur indéfinie', submitted in 1812 to the Académie Française des Sciences, for which he was awarded the 'Prix d'analyse mathématique' for 1815. Here he extensively applied Fourier transforms, and in one of the various appendices (covering 192 pages in total!) he introduced the theory of complex functions. The only 'practical' problem he investigated in this paper was the generation of waves by a paraboloid bouncing into water.

The first analytical investigations into the waves produced by a disturbance travelling at uniform speed must obviously be attributed to Wm. Thomson in the fall of 1886. In the minutes of a session on Monday, Nov. 22nd of the Royal Society of Edinburgh (Sir Wm. Thomson, president, in the chair) we may read ...

3. On ring waves produced by throwing a stone.
   (*Proc. Roy. Soc.*, Jan. 1887).

4. On waves produced by a ship advancing uniformly into smooth water. (*Phil. Mag.*, Jan. 1887).

Hence, one might hope to find more details in that journal, co-edited by

Sir William.

And in fact, already in the October 1886 issue of *Phil. Mag.*, in an article 'On stationary waves ...', we can read

> ... and I hope to include in part II, or at all events in part
> III, to be published in December, a *complete* investigation,
> illustrated by drawings, of the beautiful pattern of waves
> produced by a ship propelled uniformly through calm deep
> water.

But part II only contains a statement: *'... if I succeed in carrying out my intentions, this series ... will end with the investigation on the wave group produced by a ship, promised at the commencement of part I ... '.* Moreover, in the January issue, he even turned all his attention to the story of a spirited horse, drawing a boat after him along the Glasgow and Ardrossan canal and discovering that when the speed exceeded some critical value, the resistance decreased considerably. It is true that this had been reported by Scott Russell some 50 years before, but now Thomson had found a theoretical explanation.

Even in the subsequent various contributions of Thomson to *Phil. Mag.* up to the end of that century we cannot detect anything related to ship waves. On the other hand, in the 1898 volume, we can find a paper by a certain J.H. Michell (obviously without a degree?), communicated by the author, entitled 'On the Wave Resistance of Ships'.

Michell formulated and solved the following boundary value problem for a velocity potential Φ:

$$\Phi_{xx} + \Phi_{yy} + \Phi_{zz} = 0 \quad \text{in the domain } z < 0, \, y \neq 0,$$

where the plane $z = 0$ corresponds to the undisturbed free surface, and the coordinate y is sidewise from the ship which advances in the $+x$ direction with speed U;

$$\Phi_{xx} + k_0 \Phi_z = 0 \quad \text{on } z = 0, \text{ with } k_0 \doteq g/U^2;$$

this is the 'linearized free surface condition' combining a dynamical and a kinematic boundary condition at the wave surface $z = \zeta(x,y)$ under certain smallness assumptions; and

$$\Phi_y = \pm f \quad \text{on } y = \pm 0,$$

where the function f (which we may interpret as a source density over some part of the vertical plane of symmetry) is related to the geometry of the ship, namely to the horizontal part of the hull tangent slope.

A fundamental assumption underlying Michell's approach is that the ship is 'thin', i.e. that f is a small quantity. Michell was well aware that *the solution is to some extent indeterminate, for we may superpose any system of free waves symmetric with respect to $y = 0$ on a particular solution. ...*

His choice of solution was ... *to make the elementary diverging waves trail aft, in other words, to satisfy the condition that the ship advances into still water.*

We may learn from this outstanding pioneering paper that for determining the waves created by a moving body, we first have to represent the body by a system of singularities, then determine the associated velocity potential Φ at the upper boundary $z = 0$ of the lower half space, and then finally take the x-derivative.

Michell calculated the wave resistance just by integrating the linearised pressure $\rho U \Phi_x$ over the hull surface rather than from the wave field behind the ship. Thus he saved calculations of the wave elevation $\zeta(x, y) = \Phi_x(x, y, 0) \, U/g$. He rather referred to a lecture of Thomson's 'On ship waves', held before the Institution of Mechanical Engineers in Edinburgh on August 3rd 1887, which was obviously familiar to the Melbourne citizen Michell.

In case you succeed in locating the proceedings of this lecture, covering more than 30 pages, you may enjoy a very transparent introduction to the phenomenon of group velocity and its influence on wave resistance under the utmost avoidance of mathematics, even for the display of results. Thomson rather presented a model of the wave pattern moulded from clay. Still we may detect and extract the following findings:

(i) All wave phenomena are restricted to a domain between two lines drawn from the ship's bow and inclined to the wake on its two sides at equal angles of 19° 28′.

(ii) Theoretically, there is a disturbance to an infinite distance ahead and in any direction, but the amount of that disturbance practically is exceedingly small — imperceptible indeed — until you come to the definite lines.

(iii) There is a perfect cusp at the points of these lines.

(iv) The law of echelon for the oblique wave pattern is illustrated by *two* sets of curves which can be described through algebraic equations.[1]

(v) A diagram of the curves calculated is displayed.

(vi) Another formula, which need not be reproduced here, gives a wave height for every point.

(vii) The theoretical formula gives infinite wave height at the cusps; but this is only a theoretical supposition, though giving an interesting example of mathematical 'infinity'. Blur it, or smooth it down.

Thomson does by no means explain *why* that angle is just 19° 28′. Rather, he gives clear instructions on how to *construct* this angle by first drawing

---

[1] Here we find the only equations of this paper. Note that Thomson does *not* refer here to 'curves of constant phase'.

a circle and then a tangent from a point at one diameter's distance from the periphery.

We can only speculate about the way of Thomson's original derivations. However, one must conclude that he had obtained a single integral over 'free waves' behind the disturbance, which he then evaluated asymptotically by Stokes's method of 'stationary phase', in the version refined by Thomson.

Obviously, this lecture found worldwide distribution when incorporated into his 'Popular Lectures and Addresses' in 1891. Thus it was quoted in Lamb's 3rd edition of 'Hydrodynamics' with a footnote: 'See Sir Wm. Thomson.... See also R.E. Froude....'; in later editions, up to the last (6th), this note is supplemented by the remark: 'The investigation there referred to, based apparently on group velocity, was never published'.

However, in July 1905, at the remarkable age of 81, Sir Wm. Thomson, then raised to the peerage as Lord Kelvin, presented a rather complete derivation of his results, enriched by three more observations:

(viii) Between the straight boundaries, the amplitudes have a decay as the inverse square root of the distance R from the disturbance.

(ix) Along these 'cusp lines', there occurs a phase jump of amount $\pi/2$ between the two principal wave systems.

(x) From a table of calculated amplitudes, we may detect that the value ∞ is assigned to the divergent wave with $\theta = \pi/2$.

We should note here that the phase jump mentioned is by no means of academic interest only; it implies that extrema from the two wave systems accumulate along a line of 15° initial inclination, approaching 19° 28' against the ship's course only asymptotically. We should mention also that Hogner (1923a) has shown, using a higher order asymptotic approximation, that the phase lag varies continuously from $\pi/2$ for $\alpha = 0$ to $\pi/3$ at the cusp lines, $\alpha = \pm 19° 28'$; this makes the wave elevation continuous even across the cusp lines once the amplitude is modified to a finite value there under this approach.

In his broad presentation[2] Thomson does not include any reference to Michell's work. One may wonder if he could not have prevented the total disregard of that fundamental paper until 1923, when it was obviously re-discovered by Havelock (cf. Tuck 1989); but there is some indication that neither Lamb nor Havelock studied that late paper of Kelvin's before it was incorporated into his 'Mathematical and Physical Papers' in

---

[2]cf. *Phil. Mag.* Jan. 1904, June 05, Jan. 06 and Jan. 07, reprinted from previous *Proc. Roy. Soc. Edinb.* Note that in the conclusion of this series, i.e. in Lord Kelvin's last paper on water waves, we find an announcement: *... We hope also to apply (139) of the present paper to the fulfillment of my old promise to deal with the beautifully varying procession seen circling outwards from the place of a stone thrown into deep water....*

1910[3]; they both just refer to that popular lecture which is devoid of mathematics.[4]

Rather than repeating Kelvin's picturesque reasoning[5] let us confirm his results from an advanced point of view, as worked out by Wehausen (see Wehausen & Laitone 1960). For the wave elevation ζ due to a pressure point of dimensionless intensity H located at the coordinate origin, advancing with uniform speed U in the $+x$ direction, there is a *formal* double Fourier integral representation in polar coordinates

$$\zeta(R,\alpha) = \lim_{\mu \to 0+} \frac{1}{k_0^3 \pi} \int_{-\pi}^{\pi} \int_0^{\infty} H \frac{k^2 e^{i\psi}}{k - k_0 \sec^2 \theta + i\mu k \sec \theta} \, dk \, d\theta \quad (8.1)$$

with

$$x = R \cos \alpha, \qquad y = R \sin \alpha, \qquad \psi \doteq kR \cos(\theta - |\alpha|).$$

The above expression may equivalently be given as

$$\zeta = \Re \frac{2}{k_0^3 \pi} \text{p.v.} \int_0^{\infty} \int_{-\pi/2}^{\pi/2} H \frac{k^2 e^{i\psi}}{k - k_0 \sec^2 \theta} \, d\theta \, dk$$
$$+ \Re \frac{2}{k_0} \int_{-\pi/2}^{\pi/2} H \sec^4 \theta \, e^{i\phi} \, d\theta \quad (8.2)$$

with

$$\phi \doteq k_0 R \sec^2 \theta \cos(\theta - |\alpha|) + \pi/2$$

Here p.v. stands for Cauchy principal value, so that the double integral represents an *even* function of x; hence the additional free-wave single integral makes sure that the free waves from the double integral ahead of the disturbance are eliminated, whereas those behind the disturbance are reinforced. Our confinement to a concentrated pressure essentially simplifies our formal analysis. One should be aware however that the formal divergence of some integrals vanishes, once we consider singularity

[3]Note that Havelock (1905) published a paper in *Phil. Mag.* in November, i.e. two months earlier! But in his Royal Society paper (Havelock 1908) he wrote 'The law of amplitude is not stated by Lord Kelvin: as Prof. Lamb conjectures, his results *seem* to have been obtained by an application of the idea of group velocity (H. Lamb, 'Hydrodynamics' pp. 406/7).'

[4]It is surprising that in the 1907 translation of Lamb's 3rd edition by Johannes Friedel, Kelvin's 1905 paper *is* quoted; not only is the phase jump explained, but in a footnote it is mentioned that this had been found independently by Ekman (1906), and even a reference to Michell's paper is made!

[5]Kelvin worked with a weighted linear superposition of oblique two-dimensional solutions due to 'forcives', using mainly geometrical arguments; his choice of a constant rather than cosine weight means that his calculations do not relate to a 'pressure point', i.e. to the limit of zero submergence of an x-oriented doublet.

distributions, where H has to be considered as a complex function of the integration variable(s) with proper decay at the limits of integration.

The above expressions can be evaluated asymptotically for large distance R by a theorem for principal value Fourier integrals quoted by Wehausen & Laitone (1960) and the method of stationary phase, which we apply in a neat version essentially following Lighthill (1965).

8.2 The rule of stationary phase
Consider

$$I \doteq \int f(x)e^{ith(x)}\, dx$$

with

$$h(x) = h(x_0) + g(x), \qquad g(x) \neq 0 \quad \text{for} \quad x \neq x_0$$

and, in the interval of integration,

$$g(x) = g_0(x - x_0)^n + g_1(x), \qquad g_1 = o((x - x_0)^n) \quad \text{as } x \to x_0,$$

with $n \geq 1$. We are interested in the behaviour of I for large values of the parameter t. Taking $tg_0(x-x_0)^n$ as new integration variable, extending the integration range to infinity, we obtain with some irrelevant modifications of the integrand

$$I = f(x_0)\frac{e^{ith(x_0)}}{n|g_0 t|^{1/n}}I_n + O(1/t) \qquad \text{as } t \to \infty \tag{8.3}$$

with

$$I_n \doteq 2\int_0^\infty \frac{e^{iu\,\text{sgn}\,(g_0 t)}}{u^{1-1/n}}\, du = 2\Gamma(1/n)e^{\frac{\pi i}{2n}\,\text{sgn}\,(g_0 t)}$$

for n even, and

$$I_n \doteq 2\int_0^\infty \frac{e^{iu}}{u^{1-1/n}}\, du\,\text{sgn}\,(g_0 t) = 2\Gamma(1/n)\cos(\frac{\pi}{2n})\,\text{sgn}\,(g_0 t)$$

for n odd.

It is obvious that if the phase h becomes stationary just at an end-point of the interval of integration, we have to take only one half of the terms on the right-hand side of (8.3).

In our case, where k will stand for x, and R, x or y will stand for t, we have only two simple zeros for the derivatives of the phase (i.e. $n = 2$) corresponding to k_1 and k_2, which coincide (i.e. $n = 3$) along the cusp lines. Note that if one extremum of the phase is a maximum, the other must of necessity be a minimum, hence the sign of g_0 must differ in the two cases!

8.3   Kelvin's approximation through single wave components

The expression (8.1) can be evaluated asymptotically for large R, i.e. far away from the pressure point, by the above rule of stationary phase (or, more elegantly, by Wehausen's procedure). A first application to the double integral yields

$$\zeta(x,y) = \Re \frac{4}{k_0} \int_{-\pi/2}^{|\alpha|-\pi/2} H \sec^4 \theta \; e^{i\phi} \, d\theta + O(1/R) \tag{8.4}$$

After another application of the rule, now to this single integral, three cases have to be distinguished:

(i) For $0 < |\alpha| < \pi - \alpha_K$, where $\alpha_K = \arcsin(1/3)$, we obtain

$$\zeta = O(1/R) \tag{8.5}$$

(ii) For $\pi - \alpha_K < |\alpha| \leq \pi$ we have

$$\zeta = \frac{4}{k_0} \sqrt{\frac{2\pi}{k_0 R}} \frac{A}{\sqrt[4]{1 - 9\sin^2 \alpha}} + O(1/R), \tag{8.6}$$

where

$$A \doteq \Re \sum_{j=1}^{2} H \sec^{5/2} \theta_j \; e^{i(k_j R \cos(\theta_j - |\alpha|) + (2j-1)\pi/4)},$$

$k_j \doteq k_0 \sec^2 \theta_j$ and θ_j is given through

$$\tan \theta_j = -\frac{1}{4} \left(1 + (-1)^j \sqrt{1 - 8\tan^2 \alpha}\right) \cot \alpha; \tag{8.7}$$

we identify $j = 1$ with 'transverse waves' and $j = 2$ with 'divergent waves'. Note that we can have ϕ stationary only if the sign of $\tan |\alpha|$ is opposite that of $\tan \theta$, hence θ is non-negative for $x < 0$ and is non-positive for $x > 0$.

(iii) For $|\alpha| = \pi - \alpha_K$, we find that

$$\zeta = \frac{4}{k_0} \frac{3^{5/6}}{\sqrt[3]{k_0 R}} \Gamma(\frac{1}{3}) A_K + O(1/R)$$

where

$$A_K \doteq \Re H \sec \theta_K \; e^{ik_0 R \sec^2 \theta_K \cos(\theta_K - \alpha) + i\pi/2}$$

and

$$\theta_K \doteq \arctan(1/\sqrt{2}) = \arccos \sqrt{2/3}.$$

Thus (iii) differs in the limits regarding α both from (i) and from (ii). With (i) and (ii) we have confirmed Kelvin's findings.

Over the last century, clarifying contributions on this subject have been made from all over the world, as documented in an interesting historical survey by A. Gamst (1979), who by use of functional analysis finally found a sufficient *and necessary* far-field condition for uniqueness of the wave pattern. Here we find a record of earlier attempts and failures to formulate far-field conditions and of the continuing disregard of the phase jump phenomenon again and again (starting with Havelock (1908)) up to recent years.

I would like to give special mention here to M. Bessho (1964) in Japan, who succeeded within a terse ingenious analysis to convert the double integral representation into a compact version of a single integral along contours in a certain complex plane[6]; see §8.8 below.

Prof. Fritz Ursell has actively participated in the efforts to unveil the details and display the internal structure of the pressure point wave pattern:

(i) For divergent waves, for which the wavelength tends to zero, but the amplitude tends to infinity in the wake of the disturbance along the positive x-axis, he showed how the singular limit depends on the direction of approach from within the lower three-dimensional half space.

(ii) He derived a new uniform asymptotic representation of the wave pattern on and near the cusp lines in terms of Airy functions, not only removing the virtually singular behaviour predicted through the stationary phase approach, but also achieving continuity across the cusp lines (Ursell 1960a).

(iii) He extended Kelvin's approach to the case of waves on a slightly non-uniform flow (Ursell 1960b).

(iv) He introduced the evanescent modes supplementing the free waves in a tank of finite depth (discussion to Sharma 1963).

(v) He finally succeeded in justifying Bessho's results by re-deriving them under standards of high rigour (Ursell 1984 and 1988).

It was at the International Seminar on Wave Resistance Theory in Ann Arbor, Michigan in 1963 that I first met Prof. Ursell. I had presented a short note (Eggers 1963) entitled 'On free waves', and in the subsequent discussion, he raised a significant question:

'Is it really pertinent to decompose a wave field into a local and a far-field component?'

This touches the essence of my present lecture.

[6]Meanwhile, Bessho (1977) has given a similar representation for the more general case of a pulsating source under steady speed of advance, which still awaits evaluation under consideration of similar work by Simmgen (1968).

It is certainly true — as I shall show in more detail — that far away from a disturbance the wave can be described through (various) simple expressions. Thus it is suggestive to assign a special name to the component represented by such an expression. However, near the disturbance, none of these expressions becomes small compared to the total wave; they even become singular, or at least demand more effort for their evaluation than the residual component!

But it is essential on the other hand for a rational development of 'wave analysis' methods, i.e. for determining the wave resistance of a ship from the geometry of her wave pattern, that we can *find* a decomposition (and *different* ones for different methods) of the total velocity potential Φ and the wave elevation ζ into a (non-global!) 'main' component (with superscript m) and a 'local' (more precisely, 'non-far-field') component (with superscript l),

$$\Phi = \Phi^m + \Phi^l \qquad \text{and} \qquad \zeta = \zeta^m + \zeta^l.$$

Ideally, this decomposition should meet the following requirements:

(i) The analytical expressions for Φ^m and ζ^m should be simpler than those for Φ and ζ, respectively; in particular, they should not involve any two-fold integrals in wave numbers.

(ii) The analytical form of the main component should be such as to allow the determination of basic wave characteristics such as the wave resistance by an application of simple numerical procedures (such as taking a Fourier transform) to the associated wave pattern.

(iii) Φ^m should be a good numerical approximation to Φ, so that the deviation is everywhere small compared to its maxima in the area under observation.

(iv) The method of decomposition should be independent of the position of the ship relative to the area under observation. Insofar as there is a dependence on the relative location of the ship, the wave characteristics should come out as invariant under changes of position.

Let us consider in the sequel some examples of such decompositions.

8.4 Contour integration in the complex k-plane using polar coordinates, with k as the inner integration variable

Equation (8.1) may equivalently be written as

$$
\zeta(R, \alpha) = \frac{1}{k_0^3 \pi} \int_{-\pi/2}^{\pi/2} \left\{ \int_{C_1} H \frac{k^2 \, e^{i\psi}}{k - k_0 \sec^2 \theta} \, dk \right.
$$
$$
\left. + \int_{C_2} H \frac{k^2 \, e^{-i\psi}}{k - k_0 \sec^2 \theta} \, dk \right\} d\theta,
$$

where the contours C_1 and C_2 are shown in figure 8.1. (The analysis for contour integration with θ as the inner integration variable has been elaborated by Smorodin (1972), leading to a representation in terms of cylindrical harmonics around the vertical axis.) A transformation of C_1 and C_2 to the imaginary k-axis leads to

$$
\zeta(R, \alpha) = \Re \frac{2}{k_0^3 \pi} \int_{\alpha-\pi/2}^{\alpha+\pi/2} \int_0^\infty \frac{H k^2}{k + i k_0 \sec^2 \theta} e^{-kR|\cos(\theta-\alpha)|} \, dk \, d\theta
$$
$$
+ \Re \frac{4}{k_0} \int_{-\pi/2}^{|\alpha|-\pi/2} H \sec^4 \theta \, e^{i\phi} \, d\theta \tag{8.8}
$$

Hence for $R \to \infty$, we find

$$
\zeta(R, \alpha) = \Re \frac{4}{k_0} \int_{-\pi/2}^{|\alpha|-\pi/2} H \sec^4 \theta \, e^{i\phi} \, d\theta + O(1/R) \tag{8.9}
$$

which represents the contribution from the residues. We have mentioned earlier that this single integral could as well have been found from (8.2) by the stationary phase rule. The integral (8.9) has been used as ζ^m for some earlier methods of wave analysis. But through the variable integration limit and dependence on R, it shows an unpleasant dependence on the choice of coordinate origin.

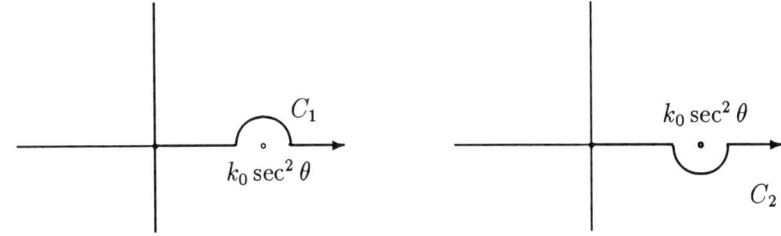

Figure 8.1: Contours in the complex k-plane.

8.5   Contour integration in the complex w-plane using Cartesian coordinates with w as the inner integration variable

Passing to Cartesian coordinates in the Fourier space leads to

$$\zeta(x,y) = \frac{1}{k_0^3 \pi} \int_{-\infty}^{\infty} \int_{-\infty}^{\infty} H \frac{w^2 e^{i\psi}}{w^2 - k_0 \sqrt{u^2 + w^2}} \, dw \, du$$

with

$$\psi \doteq wx + u|y|.$$

Through proper selection of the contour C_3 for the complex w-integration (see figure 8.2), we can ensure that all waves disappear for $x \to +\infty$. A transformation of the contour around the branch cuts on the imaginary axis (see figure 8.2) then leads to

$$\begin{aligned}
\zeta &= -\Re \frac{2}{k_0^3 \pi} \int_{-\infty}^{\infty} \int_{|u|}^{\infty} H \frac{w^2 e^{-w|x|+iuy}}{k_0 \sqrt{w^2 - u^2} - iw^2} \, dw \, du \\
&\quad + \Re \frac{4}{k_0^2} \frac{1 - \operatorname{sgn}(x)}{2} \int_{-\infty}^{\infty} H \frac{s^3(u) e^{i\phi}}{\sqrt{1 + 4u^2/k_0^2}} \, du
\end{aligned} \qquad (8.10)$$

with

$$s(u) \doteq \sqrt{\left(1 + \sqrt{1 + 4u^2/k_0^2}\right) / 2}$$

and

$$\phi \doteq k_0 s(u) x + u|y| + \pi/2.$$

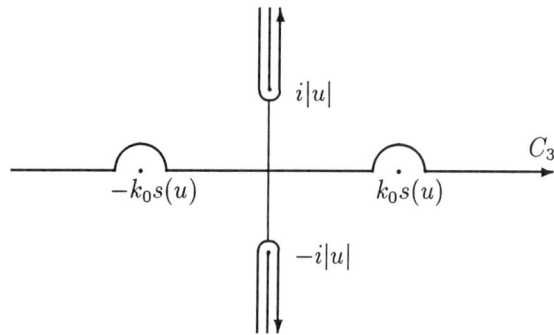

Figure 8.2: Contour C_3 in the complex w-plane and its transformation around the branch cuts (upper one for $x > 0$, lower one for $x < 0$).

For $x \to -\infty$ we find

$$\zeta(R, \alpha) = \Re \frac{4}{k_0} \int_{-\pi/2}^{\pi/2} H \sec^4 \theta \, e^{i\phi} \, d\theta + O(1/x) \tag{8.11}$$

again as the contribution from the residuals, valid for *all* y i.e. for *all* α.

Taking the single integral term from (8.11) as ζ^m, with H as an arbitrary complex function of u, one can derive a method for determining the wave resistance of a ship from the wave profiles of (at least) two transverse cuts ($x = $ constant) across the wake.

The above decomposition has the major advantage that it can easily be extended to the case of a pressure point (or more generally of a singularity system) amidst a tank of finite width b. Assuming vertical walls at $y = \pm b/2$, through considering an infinite echelon of images, we obtain through Poisson's summation rule (exploiting the fact that the Fourier transform of a row of equidistant δ-functions is such a row in the Fourier space (see Lighthill 1958))

$$\zeta^m = \frac{1 - \text{sgn}\,(x)}{k_0^2 \pi b} \Re \sum_{\nu=-\infty}^{\infty} H(u_\nu) \frac{s_\nu^3}{\sqrt{1 + 4(u_\nu/k_0)^2}} e^{i(k_0 s_\nu x + u_\nu y + \pi/2)}$$

with $u_\nu \doteq \nu\pi/b$ and $s_\nu \doteq s(u_\nu)$. Note that this is just the expression for evaluating the integral in (8.11) by the trapezoidal rule with a step width of $\Delta u = 1/(\pi b)$.

ζ^m thus defined is the basis for performing wave analysis in a tank of width b with wave reflection at the vertical side walls $y = \pm b/2$ (Eggers *et al.* 1968). The wave resistance can be expressed as a weighted sum of the squared absolute values of the y-Fourier coefficients $H(u_\nu) \, e^{ik_0 s_\nu x_0}$ at any cut $x = x_0$ behind the ship. The wave elevation for at least two transverse cuts must be measured.

The pertinence of modelling the wave pattern by the approximation ζ^m has been confirmed by Sharma (1963) measuring simultaneously 27 cuts, each with 128 equidistant data points from stereometric evaluation of photographs from above!

This approximation is valuable even for a wide tank with practically no reflection, if we operate with a virtual tank width which exceeds that part of the cuts where non-zero data are taken.

8.6 Contour integration in the complex u-plane using Cartesian coordinates with u as the inner integration variable

Considering again

$$\zeta(x, y) = \frac{1}{k_0^3 \pi} \int_{-\infty}^{\infty} \int_C H \frac{w^2 e^{i\psi}}{w^2 - k_0 \sqrt{u^2 + w^2}} \, du \, dw$$

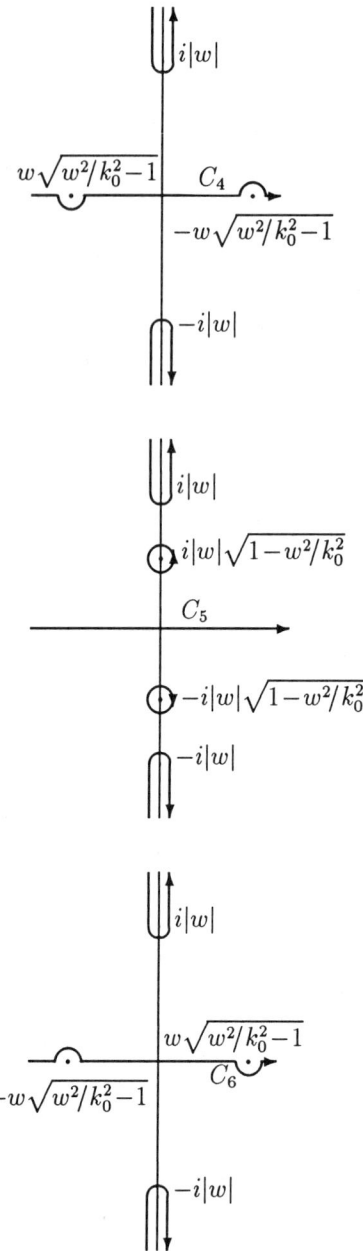

Figure 8.3: Contours in the complex u-plane and their transformations around the branch cuts (upper ones for $wy > 0$, lower ones for $wy < 0$). Use C_4 for $w < -k_0$, C_5 for $-k_0 < w < k_0$, and C_6 for $w > k_0$.

we can satisfy our radiation condition, if we select the contour C in the complex u-plane as shown in figure 8.3. A transformation of the contours along branch cuts on the imaginary axis then leads to

$$
\begin{aligned}
\zeta \;=\;& \Re \frac{2}{k_0^3 \pi} \int_{-\infty}^{\infty} \int_{|w|}^{\infty} H \frac{w^2 e^{iwx - u|y|}}{k_0 \sqrt{u^2 - w^2} - iw^2} \, du \, dw \\
&+ \Re \frac{4}{k_0^4} \int_0^{k_0} H \frac{w^3 e^{iwx - w\sqrt{k_0^2 - w^2}\,|y|/k_0}}{\sqrt{k_0^2 - w^2}} \, dw \\
&+ \Re \frac{4}{k_0^4} \int_{k_0}^{\infty} H \frac{w^3 e^{iwx + iw\sqrt{w^2 - k_0^2}\,|y|/k_0 + i\pi/2}}{\sqrt{w^2 - k_0^2}} \, dw
\end{aligned}
\tag{8.12}
$$

where again the single integrals represent the residue contributions. Passing to the limit $|y| \to \infty$, we obtain

$$
\zeta(R, \alpha) = \Re \frac{4}{k_0} \int_0^{\pi/2} H \sec^4 \theta \, e^{i\phi} \, d\theta + O(1/y)
\tag{8.13}
$$

in our earlier polar notation, valid for *all* x. Starting with ζ^m as the last line of (8.12), with H an arbitrary complex function of θ, one can derive a method for determining the wave resistance of a ship from one wave profile (at least) along longitudinal cuts $y = $ constant on either side parallel to the course of the ship; in symmetric cases, one can even do with one side only. The single integral with exponential decay in y contains a spectral range with no contribution to wave resistance; its role is obviously to provide continuity of the stationary phase limit with y approaching zero for $x < 0$.

8.7   Discussion of the different decompositions

We should mention here that even the far-field expressions derived by Kelvin, i.e. (8.4), (8.5), and (8.6), define a decomposition; it has actually been exploited in primary stages of wave analysis by Kajitani (1963). But it is evident that this decomposition is sensitive to a change of the coordinate origin for which there is ample variance in case of ship waves.

Observe now the vexing result that for an arbitrary point the first three decompositions lead to single integral representations of ζ^m (namely (8.9), (8.11) and (8.13)) with different integration limits when expressed as θ-integrals for x and y given! One may certainly ask which then gives the 'best' approximation, and there is no general answer, unless we have decided for a wave analysis method, where ζ^m has to be integrated over an infinite range in general. So far, even other θ integration ranges could be admitted to approximate ζ, provided they include those (positive) values θ_1, θ_2 for which the phase becomes stationary; this is sufficient to obtain identical far-field expressions. Here Bessho's single-integral representation of the total wave elevation can give us some insight for an assessment of the alternative results.

8.8   Bessho's compact single-integral representation of the wave field

Bessho's representation (1964) for ζ is

$$-\frac{\partial^2}{\partial x^2} \Re \frac{2}{k_0^3} \int_{L_1+L_2} H \sinh v \; e^{-k_0 \sinh v \, (i|y| \cosh v - x)} \, dv$$

where the contours L_1 and L_2 are shown in figure 8.4

(we have substituted $-x$ for x, and $|y|$ for $-y$, as ζ is even in y). In order to obtain a factor $e^{i\phi}$, we insert our wave front angle θ through

$$i \cosh v = \tan \theta \Rightarrow \sinh v = i \cosh(v - i\pi/2) = i \sec \theta$$

for the horizontal part of L_2,

$$i \cosh v = \tan \theta \Rightarrow \sinh v = -i \cosh(v + i\pi/2) = -i \sec \theta$$

for L_1 and finally, for the vertical part of L_2,

$$\cosh v = -w \Rightarrow i \sinh v = \sqrt{1 - w^2}.$$

Then we obtain (in accord with the investigations of H.T. Wang (1989), who simplified Ursell's analysis (1984))

$$
\begin{aligned}
\zeta(x,y) \;=\;& \Re \frac{1}{2} \left[\frac{4}{k_0} \int_0^{\pi/2} H \sec^4 \theta \; e^{i\phi} \, d\theta \right. \\
&+ \left. \frac{4}{k_0^4} \int_0^{k_0} H \frac{w^3 e^{iwx + w\sqrt{k_0^2 - w^2}\,|y|/k_0}}{\sqrt{k_0^2 - w^2}} \, dw \right] \\
&+ \Re \frac{1}{2} \left[\frac{4}{k_0} \int_{-\pi/2}^0 H \sec^4 \theta \; e^{i\phi} \, d\theta \right. \\
&+ \left. \frac{4}{k_0} \int_0^{|\alpha| - \pi/2} H \sec^4 \theta \; e^{i\phi} \, d\theta \right] \\
&+ \frac{2}{k_0^2} H \frac{R}{y^2}
\end{aligned}
$$

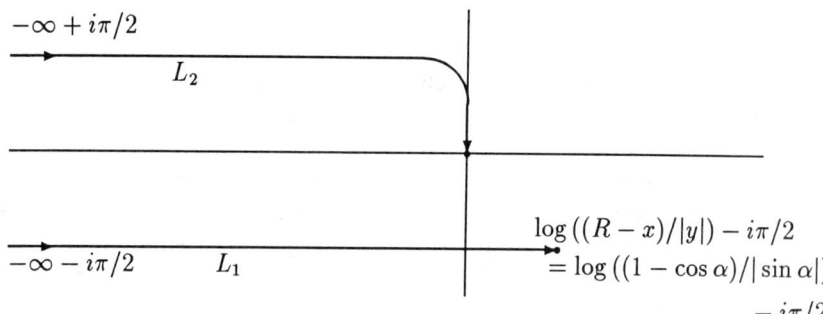

$-\infty + i\pi/2$

L_2

$-\infty - i\pi/2$ L_1

$\log\left((R - x)/|y|\right) - i\pi/2$
$= \log\left((1 - \cos\alpha)/|\sin\alpha|\right)$
$\qquad - i\pi/2$

Figure 8.4: Bessho's contours in a complex v-plane.

We may thus observe that — save a rational expression from differentiating a variable limit integral with respect to x — Bessho's expression is an average of the limit $|y| \to \infty$ of (8.12) — the contribution from L_2 — and the limit $R \to \infty$ of (8.10) — the contribution from L_1 — which contains as well half of the limit $x \to -\infty$ of (8.8) represented by the first and third term! Note that the combination of these four terms ensures that, prior to the twofold x-differentiation, an asymptotic evaluation remains continuous in the polar angle even for α approaching 0 and π i.e. $y = 0$; the second term balances the step of the first in the amplitude for $\theta = \theta_1$; the fourth term cancels any asymptotic contribution from the third for $x > 0$ i.e. $|\alpha| < \pi/2$.

From the above expression we may be inclined to define another decomposition, where only the rational term obtained from the influence of the variable integration limit on the differentiation (not present in the related expression for the source potential itself as given by Bessho and investigated by Ursell (1984)) is left to define the local component ζ', but we should observe that this term is infinite along the entire x-axis. As such singular behaviour should not occur for $x > 0$ ahead of the pressure point, we may expect that it is compensated (at least for $x > 0$) by the wave integral terms which are singular there as well, so that finally Ursell's investigations (1960b) can be confirmed. Let us hence study the limit from out of the entire three-dimensional space, for which Bessho's expression reads

$$U\Phi_x(x,y,z)/g = -\frac{\partial^2}{\partial x^2} \Re \frac{2}{k_0^3} \int_{L_1+L_2} H \sinh v \, e^{\mathcal{E}(v)} \, dv,$$

where

$$\mathcal{E}(v) \doteq k_0 \rho (\sinh \beta + i \cosh(v + i\gamma)) \sinh v,$$

$\rho \doteq \sqrt{y^2 + z^2}$, $\beta \doteq \log(\frac{R-x}{\rho})$ and $\gamma \doteq \arctan(|y/z|)$; the variable integration limit for the contour L_1 now has to be taken as $\beta - i\gamma$. With the above substitutions we then need the limit $\rho \to 0$ of

$$
\begin{aligned}
U\Phi_x/g &= \Re \frac{1}{2} \left[\frac{4}{k_0} \int_0^{\pi/2} H \sec^4 \theta \, e^{i\phi + k_0 z \sec^2 \theta} \, d\theta \right. \\
&\quad + \frac{4}{k_0^4} \int_0^{k_0} H \frac{w^3 e^{iwx + w\sqrt{k_0^2 - w^2}|y|/k_0 + w^2 z/k_0}}{\sqrt{k_0^2 - w^2}} \, dw \left. \right] \\
&\quad + \Re \frac{1}{2} \left[\frac{4}{k_0} \int_{-\pi/2}^0 H \sec^4 \theta \, e^{i\phi + k_0 z \sec^2 \theta} \, d\theta \right. \\
&\quad + \frac{4}{k_0} \int_0^{|\alpha| - \pi/2} H \sec^4 \theta \, e^{i\phi + k_0 z \sec^2 \theta} \, d\theta \left. \right] \\
&\quad + \frac{2H}{k_0^2} \left[\frac{Ry^2}{\rho^4} - \frac{x^2 z^2}{\rho^4 R} - \frac{z}{k_0 R^3} \right]
\end{aligned}
\tag{8.14}
$$

where $R = \sqrt{x^2 + \rho^2}$. Here, we have noted that

$$\frac{\partial \beta}{\partial x} = -\frac{1}{R}, \qquad \frac{\partial}{\partial x}\left(\frac{xz}{\rho^2 R}\right) = \frac{z}{R^3}$$

and that at the variable integration limit the integrand is

$$\sinh(\beta - i\gamma) = -\frac{xz}{\rho^2} + i\frac{|y|R}{\rho^2}.$$

Obviously, at least for the last line of (8.14), the limit $\rho \to 0$ depends on γ; we should observe that (save the factor $2H/k_0^3$) the last term reads $(\partial/\partial z)R^{-1}$ and approaches $2\pi\delta(x,y)$ i.e. a Dirac δ-function in the plane $z = 0-$, corresponding to a vertical doublet in the terminology of physics, as first observed by Brard (1974); note that this cannot be conceived as representing a downward flux.

With ρ approaching zero, three of the above integrals become divergent, and the rational expression becomes singular for any γ. Assuming, however, that this limit process commutes with twofold x-differentiation, for constant H at least, we can obtain a closed-form representation which is finite for all $x \neq 0$.

Introducing the dimensionless variables

$$\mathcal{X} \doteq k_0 x, \quad \mathcal{R} \doteq k_0 \rho \quad \text{and} \quad \mathcal{Z} \doteq -k_0 z,$$

we obtain

$$\zeta(x,0) = \frac{H}{k_0}\left[2 + \frac{2}{|\mathcal{X}|} - \pi(\mathbf{H}_1(\mathcal{X}) - Y_1(|\mathcal{X}|))\right]''$$
$$+ \frac{H}{k_0}\frac{1 - \mathrm{sgn}(\mathcal{X})}{2}[\pi Y_1(|\mathcal{X}|)]''$$

where $Y_n(x)$ is a Bessel function and $\mathbf{H}_n(x)$ is a Struve function (even for n odd). Note that the first term is singular of order $\mathcal{X}^{-1}$ only, whereas the second term is singular of order $\mathcal{X}^{-3}$ and represents the outgoing waves.

Gamst has demonstrated that for the entire domain $\mathcal{R} \neq 0, 0 \leq \gamma \leq 2\pi$, Bessho's contour integral representation can be made convergent through a γ-dependent vertical shift of the contour initial points at infinity, thus defining a single-valued function in the entire three-dimensional space around the x-axis. For the outgoing waves, related to the second term of (8.10), this could have been anticipated from a representation given by Hogner (1923b) for $\mathcal{R} > 0$ and $\mathcal{X} < 0$, equivalent to

$$\Re \int_{-\pi/2}^{\pi/2} \sec^4 \theta \, e^{i\phi - \mathcal{Z}\sec^2 \theta} \, d\theta = \frac{1}{2}\sum_{n=1}^{\infty} \frac{\mathcal{X}^{2n-1}}{(2n-1)!}\frac{\partial^n E}{\partial \mathcal{Z}^n}, \qquad (8.15)$$

where

$$E \doteq e^{-\mathcal{Z}/2} \left[K_0(\mathcal{R}/2) + K_1(\mathcal{R}/2) \cos\gamma \right],$$

and $K_n(x)$ is a modified Bessel function. If we observe that $\partial \mathcal{R}/\partial \mathcal{Z} = \cos\gamma$, we could rearrange the terms and thence derive a formal Fourier series in γ, as has been done by Bessho (1964). Note also that Newman (1987) has investigated the case $\gamma = 0$ and given a representation through a Neumann series from which we can evaluate the left-hand side of (8.15) on $y = 0$, $z < 0$ as

$$e^{-\mathcal{Z}/2} \sum_{n=0}^{\infty} (-1)^n J''_{2n+1}(\mathcal{X}) \left[K_n(\mathcal{Z}/2) + K_{n+1}(\mathcal{Z}/2) \right]. \qquad (8.16)$$

Note that, as $K_n(x)$ is singular at $x = 0$, (8.15) diverges as $\mathcal{R} \to 0$ and (8.16) diverges as $\mathcal{Z} \to 0$.

The above analysis makes evident, as Ursell conjectured in his discussion (Eggers 1963), that no continuous *global* far-field free wave integral expression can be given.

8.9 The determination of wave resistance from measured wave profiles

8.9.1 Profiles along a longitudinal cut

Assuming symmetry with regard to the coordinate y, the wave resistance R_w may be considered as twice the flux of x-momentum through an infinite vertical plane $y = y_c$ parallel to the ship's course, up to the undisturbed free surface $z = 0$ (discarding the contribution from the wave elevation as a higher-order term),

$$R_w = 2\rho \int_{-\infty}^{\infty} \int_{-\infty}^{0} \Phi_x \Phi_y \, dz \, dx.$$

This expression can be evaluated in terms of ζ (or of the lengthwise wave slope ζ_x, or of the transverse slope ζ_y) under *two* assumptions for the associated velocity potential Φ near the plane $y = y_c$, namely

$$\Phi_{xx} + k_0 \Phi_z = 0$$

which is analogous to that for a downward heat flow, and the Laplace equation

$$\Phi_{xx} + \Phi_{yy} + \Phi_{zz} = 0.$$

The first assumption (which is met for Φ related to ζ^m obtained from (8.12)) implies that under very general conditions we have

$$\Phi(x, y_c, z) = \frac{1}{2} \sqrt{\frac{k_0}{\pi |z|}} \int_{-\infty}^{\infty} \Phi(x', y_c, 0) e^{k_0 (x-x')^2 / (4z)} \, dx'$$

$$= \frac{k_0}{2\pi} \int_{-\infty}^{\infty} \Phi(x', y_c, 0) \int_{-\infty}^{\infty} e^{k_0 w^2 z} \cos\left(k_0 w(x - x')\right) dw\, dx'.$$

This means that the Fourier transforms of Φ, Φ_x and Φ_y depend on z just via a factor $e^{k_0 w^2 z}$. Thus if we define

$$X(w, z) \doteq \frac{1}{2\pi} \int_{-\infty}^{\infty} \Phi_x(x, y_c, z) e^{ik_0 w x}\, dx$$

and

$$Y(w, z) \doteq \frac{1}{2\pi} \int_{-\infty}^{\infty} \Phi_y(x, y_c, z) e^{ik_0 w x}\, dx,$$

we must accordingly have

$$X(w, z) = X(w, 0) e^{k_0 w^2 z} = e^{k_0 w^2 z} \frac{g}{2\pi U} \int_{-\infty}^{\infty} \zeta(x, y_c) e^{ik_0 w x}\, dx$$

and

$$Y(w, z) = Y(w, 0) e^{k_0 w^2 z}.$$

Then we obtain from Parseval's theorem

$$
\begin{aligned}
R_w &= 2\rho \int_{-\infty}^{\infty} \int_{-\infty}^{0} \Phi_x \Phi_y\, dz\, dx \\
&= 4\pi \rho k_0 \,\Re \int_{-\infty}^{\infty} \int_{-\infty}^{0} X(w, 0) Y(-w, 0) e^{2k_0 w^2 z}\, dz\, dw \\
&= 2\pi \rho \,\Re \int_{-\infty}^{\infty} \frac{X(w, 0) Y(-w, 0)}{w^2}\, dw.
\end{aligned}
$$

At first glance it may appear that we have to determine both $X(w, 0)$ *and* $Y(w, 0)$; but from Laplace's equation (and from reasoning that R_w should be non-negative) we may deduce that

$$Y(w, 0) = \sqrt{w^2 - 1}\, X(w, 0)$$

or more generally that

$$XX(w) \doteq k_0 U \int_{-\infty}^{\infty} \zeta_x(x, y_c) e^{ik_0 w x}\, dx = i w k_0 X(w, 0)$$

and

$$XY(w) \doteq k_0 U \int_{-\infty}^{\infty} \zeta_y(x, y_c) e^{ik_0 w x}\, dx = i w k_0 \sqrt{w^2 - 1}\, X(w, 0).$$

It can be further seen that there is no contribution to R_w from the 'long-wave' range $-k_0 < k_0 w < k_0$.

For optimum application of the above result, one may ask if ζ, ζ_x or ζ_y should be the object of experimental measurements. Here the requirement that the quantity measured should have a fast decay with x (in order to keep records short and to avoid the range where tank wall influence may be felt) conflicts with the desire to have the Fourier transform intensity concentrated in the long wave range.

8.9.2   *Profiles along a transverse cut*

We consider a tank with vertical walls at $y = \pm b/2$ and depth d, and we assume that the waves are symmetric about the tank centre plane $y = 0$. From considerations of pressure integration and momentum flux, the wave resistance R_w can be expressed in terms of the flow and wave profile at a transverse cut $x = x_c$ as

$$R_w = \frac{\rho}{2} \int_{-b/2}^{b/2} \int_{-d}^{0} \{-\Phi_x(x_c, y, z)^2 + \Phi_y^2 + \Phi_z^2\} \, dz \, dy$$

$$+ \frac{\rho g}{2} \int_{-b/2}^{b/2} \zeta(x_c, y)^2 \, dy. \tag{8.17}$$

Let us imagine that the wave pattern near $x = x_c$ can be represented through a y-Fourier series, i.e. that it is composed of components

$$\zeta_\nu(x, y) = \Re \, a_\nu e^{ik_\nu x \cos \theta_\nu + iu_\nu y}$$

(where $k_\nu \doteq k_0 \sec^2 \theta_\nu$, $u_\nu \doteq k_\nu \sin \theta_\nu$ and θ_ν must satisfy $u_\nu = \nu \pi / b$ so that the tank-wall boundary conditions are satisfied) with phase velocity $c_\nu = U \cos \theta_\nu$ so that each component is stationary with respect to the ship, travelling at speed U. Then the time average of the increase of energy in the domain $x > x_c$ is $\frac{1}{2} U \rho g |a_\nu|^2 b$. On the other hand, there is an average flow of energy through the plane $x = x_c$ of amount $\frac{1}{2} U \rho g |a_\nu|^2 c_g \cos \theta_\nu$. The difference of these two energy terms must be supplied to the fluid by overcoming the wave resistance R_w.

Hence[7] if the wave elevation near $x = x_c$ can be represented by

$$\zeta(x, y) = \Re \sum_{\nu = -\infty}^{\infty} a_\nu e^{ik_\nu x \cos \theta_\nu + iu_\nu y}$$

with $a_\nu = a_{-\nu}$. We can then follow arguments of Lamb and conclude (without explicit evaluation of (8.17)) that

$$R_w = \frac{\rho g b}{2} \sum_{\nu = -\infty}^{\infty} \left(1 - \frac{c_g}{c_\nu} \cos^2 \theta_\nu\right) |a_\nu|^2.$$

The same analysis holds for antisymmetric wave patterns with $a_{1-\nu} = -a_\nu^*$ and $u_\nu = (2\nu + 1)\pi/(2b)$, and combinations of both. (a^* is the complex conjugate of a.) We can easily find therefrom the formula for infinite tankwidth ($b \to \infty$): replace

$$a_\nu \quad \text{by} \quad H \frac{s^3}{\sqrt{1 + 4(u/k_0)^2}} \frac{\pi}{k_0^2 b},$$

π/b by du and the summation over ν by integration over u.

---

[7]The occurrence of additional, evanescent wave modes in the near field — by analogy with the wavemaker problem — and their rate of decay (as later displayed by Wehausen 1972) is elucidated in Ursell's discussion to Sharma (1963).

8.10   The ray approach to the ship-wave problem

From a comparison of a term $e^{i(sx+uy)}$ (where $s = s(u)$) with a time-dependent wave term $e^{i(kx-\omega t)}$ (where $k = k(\omega)$), we observe a direct analogy between the non-stationary wave problem and the stationary ship-wave problem, as indicated by M.J. Lighthill (1967). It is suggestive then to define 'rays' in the x-y plane through their slope

$$\frac{dy}{dx} = -\frac{ds}{du},$$

where the right-hand side is analogous to the group velocity $d\omega/dk$ in the time-dependent case. In our example of Kelvin's wave pattern, such rays are straight lines through the origin.

Ursell (1960b) has generalized this concept to the case of stationary waves over an inhomogeneous slowly varying basic flow, rather than uniform parallel flow as underlying our analysis so far. It is then pertinent to measure the wave normal angle θ against the local flow direction and consider the 'dispersion relation' for the associated wave number k, governed by the magnitude of the *local* flow. We find that the ray direction is the resultant of the basic flow and some 'local group velocity vector' derived from a *local* phase velocity c, which balances the basic flow component normal to the wave front; this is equivalent to Lighthill's definition of the ray direction through

$$\frac{dy}{dx} = -\frac{\partial s}{\partial u},$$

where the right-hand side replaces the group velocity and is derived from the 'dispersion relation' $s = s(u, x, y)$.

For the ray direction angle α, we can then obtain the dependence on the wave normal angle θ (both measured against the basic flow direction) through

$$\tan \alpha = \frac{c_g \sin \theta}{c_g \cos \theta - Uq} = \frac{\sin 2\theta}{1 + \cos 2\theta - 2c/c_g},$$

where Uq stands for the local flow intensity. This generalizes Kelvin's relation (8.7) between α and θ which led to the stationary values θ_1 and θ_2. One may even introduce an 'action transport velocity' $c_{at} \doteq -c_g \sin \theta / \sin \alpha$ as the resultant of the basic flow and the group velocity vector. This concept provides a tool for explaining the deviation of the Kelvin wave cusp angle in ship model experiments from $19°28'$ as predicted and found in the far field. In addition, effects of surface tension and of finite water depth can be taken care of (see Eggers & Huang (1990) and Eggers (1981)).

For such inhomogeneous flows, there is little reason to assume that both θ *and* k (or u *and* s) remain constant along a ray, as the dispersion relation now is dependent on the space coordinates. Thus we need some conservation principle, which hopefully may serve us even near the ship. By postulating 'conservation of wave number' or equivalently asking for irrotationality of the wave number vector, the determination of rays becomes equivalent to constructing characteristics for a certain partial differential equation for an 'Eikonal function' $S(x, y)$ for which the wave number vector is the gradient. This differential equation certainly admits regular solutions in the entire domain where the basic flow is regular. Do these solutions still provide an adequate appoximation to the ship wave pattern there, in particular if the basic flow is no longer of slow variation (cf. Keller 1979)?

8.11   Conclusions
The single integral approximations I have reviewed here range in some hybrid position between exact expressions and consistent far field asymptotics. From a formal analysis, we may only say that they become valid far away from the disturbance. But nature was kind to the naval architect: As close as half a ship's length behind the stern, or say five times the breadth sidewise, these appoximations lend themselves to meaningful application of ship wave analysis and they make the resistance calculated practically independent of the position of the ship with regard to the domain where wave elevation data are taken, as we should require of a consistent procedure.

It should be emphasized on the other hand that such approximations (and still more a ship ray theory based on asymptotic evaluation of such approximations) cannot have uniform validity. In particular, our example of a pressure point wave pattern shows that however small the Froude number may be, we should expect the existence of some 'boundary layer' (in the spirit of singular perturbation theory) around the ship's hull where such approximations are no longer pertinent, and hence no boundary condition can be satisfied. I hope that my lecture has added to substantiating this argument.

8.12   References

Bessho, M. 1964 On the fundamental function in the theory of wavemaking resistance of ships. *Memoirs of the Defense Academy, Japan* **4**, 99–119.

Bessho, M. 1977 On the fundamental singularity in the theory of ship motions in a seaway. *Memoirs of the Defense Academy, Japan* **17**, 3, 94–105.

Brard, R. 1974 Le problème de Neumann-Kelvin. *C. R. Acad. Sci. Paris*, Ser. A **278**, 163–167.

Cauchy, A.L. 1827 Théorie de la propagation des ondes à la surface d'un fluide pesant d'une profondeur indéfinie. *Mémoires présentés par divers Savants à l'Académie royale des Sciences de l'Institut de France* **1**, 3–123, Notes 124–316. Also *Œuvres Complètes*, Série I, tome I (1882) 5–318.

Eggers, K. 1963 On free waves. *Int. Seminar on Theoretical Wave Resistance*, Ann Arbor, Mich. **1**, 191–196, disc. 197–200, disc. by F. Ursell 197.

Eggers, K., Sharma, S.D. & Ward, L.W. 1967 An assessment of experimental methods for determining the wavemaking characteristics of a ship form. *Trans. Soc. Naval Arch. and Mar. Eng.* **75**, 112–144, disc. 145–157.

Eggers, K. 1976 Wave analysis, state of the art 1975. *Proc. Int. Seminar on Wave Resistance Theory*, Tokyo, 93–106.

Eggers, K. 1981 Non-Kelvin dispersive waves around non-slender ships. *Schiffstechnik* **28**, 223–245, disc. 245–252, disc. by F. Ursell 248.

Eggers, K. & Huang, D.B. 1990 Ship wave ray tracing including surface tension. Bericht Nr. 501, Institut für Schiffbau der Univ. Hamburg.

Ekman, V.W. 1906 On stationary waves on running water. *Arkiv för matematik, astronomi och fysik* **3**, No. 2, 1–30.

Gamst, A. 1979 Existenz, Eindeutigkeit und Regularität stationärer Wellenströmungen, die von Druckverteilungen an der Wasseroberfläche verursacht werden – lineare Theorie. Bericht Nr. 391, Institut für Schiffbau der Univ. Hamburg.

Havelock, T.H. 1905 On surfaces of discontinuity in a rotationally elastic medium. *Phil. Mag.* (5) **10**, 603–612.

Havelock, T.H. 1908 The propagation of groups of waves in dispersive media, with application to waves on water produced by a travelling disturbance. *Proc. Roy. Soc.* A**82**, 398–430.

Hogner, E. 1923a Contributions to the theory of ship waves. *Arkiv för matematik, astronomi och fysik* **17**, No. 12, 1–50.

Hogner, E. 1923b Notes on some new contributions to the theory of ship waves. *Arkiv för matematik, astronomi och fysik* **18**, No. 18, 1–9.

Kajitani, H. 1963 Wave resistance obtained from photogrammetrical analysis of the wave pattern. *Int. Seminar on Theoretical Wave Resistance*, Ann Arbor, Mich. **1**, 417–450, disc. 451.

Keller, J.B. 1979 The ray theory of ship waves and the class of streamlined ships. *J. Fluid Mech.* **91**, 465–487.

Kelvin, Lord 1905 Deep sea ship-waves. *Proc. Roy. Soc. Edinb.* **25**, 1060–1084, issued separately Dec. 1905; also *Phil. Mag.* (6) **11** (1906) 1–25 and *Mathematical and Physical Papers*, Vol. IV (1910) 394–418.

Lamb, H. 1907 *Lehrbuch der Hydrodynamik*, autorisierte Übersetzung der dritten Auflage von Johannes Friedel. Leipzig-Berlin: B.G. Teubner.

Lighthill, M.J. 1958 *Introduction to Fourier Analysis and Generalized Functions*. Cambridge University Press.

Lighthill, M.J. 1965 Group velocity. *J. Inst. Maths Applics.* **1**, 1–28.

Lighthill, M.J. 1967 Special cases treated by the Whitham theory. *Proc. Roy. Soc.* A**229**, 28–53.

Michell, J.H. 1898 The wave resistance of a ship. *Phil. Mag.* (5) **45**, 106–123.

Newman, J.N. 1987 Evaluation of the wave-resistance Green function: Part 2 — The single integral on the centerplane. *J. Ship Res.* **31**, 145–150.

Sharma, S.D. 1963 A comparison of the calculated and measured free wave spectrum of an Inuid in steady motion. *Int. Seminar on Theoretical Wave Resistance*, Ann Arbor, Mich. **1**, 203–250, disc. 251–285, disc. by F. Ursell 255–270.

Simmgen, M. 1968 Ein Beitrag zur linearisierten Theorie des periodisch instationär angeströmten Unterwassertragflügels. *Z. angewandte Math. Mech.* **48**, 255–264.

Smorodin, A.I. 1972 Waves at the fluid surface during the motion of a submerged ellipsoid. *Prikl. Mat. Mekh.* **36**, 148–152.

Thomson, W. 1887 On ship waves. *Proc. Instn. Mech. Engrs.* 409–433; also *Popular Lectures and Addresses* Vol. II (1891) 450–500.

Tuck, E.O. 1989 The wave resistance formula of J.H. Michell (1898) and its significance to recent research in ship hydrodynamics. *J. Austral. Math. Soc. B* **30**, 369–377.

Ursell, F. 1960a On Kelvin's ship wave pattern. *J. Fluid Mech.* **8**, 418–431.

Ursell, F. 1960b Steady wave patterns on non uniform fluid flow. *J. Fluid Mech.* **9**, 337–364.

Ursell, F. 1984 Mathematical note on the fundamental solution (Kelvin source) in ship hydrodynamics. *IMA J. Appl. Math.* **32**, 335–351.

Ursell, F. 1989 On the theory of the Kelvin ship wave source: asymptotic expansion of an integral. *Proc. Roy. Soc.* A**418**, 81–93.

Wang, H.T. 1989 Calculation of the odd and even integral components of the wave resistance Green's function. Naval Research Laboratory Mem. Rept. 6411, 37 pp, Washington D.C.

Wehausen, J.V. 1973 The wave resistance of ships. *Adv. Appl. Mech.* **13**, 93–245.

Wehausen, J.V. & Laitone, E. 1960 Surface waves. *Handbuch der Physik*, Springer Verlag, **9**, 446–778.

9

An analytical theory of propeller-generated effective wake

John P. Breslin
Consultant

I have had the privilege of knowing Fritz Ursell since a memorable visit with Sir Geoffrey Taylor in the Cavendish Laboratory in 1960. At that time, Fritz kindly took me about the campus and discussed free-surface Green functions. He related that he had just been offered the chair in mathematics recently vacated by M.J. Lighthill at Manchester, and had sought the advice of Sir Geoffrey about the wisdom of that move. Sir Geoffrey's reply was (in effect)... 'whatever decision you make, there will be times when you regret it'! Fritz's long success at Manchester stands in contrast to that counsel of despair! (One wonders what the Eminent Master regretted?)

In the intervening years, I have sought Fritz's counsel regarding mathematical difficulties which he kindly resolved with deft strokes. It is a great honour to contribute this paper to this commemorative volume, although it does not fit the theme of 'Wave asymptotics'; perhaps it can be thought of as an asymptotic theory of viscous shear flow at infinite Reynolds number?

9.1 Introduction

Effective wake is a term applied by naval architects to distinguish the flow as 'seen' by ship propellers operating abaft a hull from the *nominal wake* deduced from measurements made abaft models in the plane of the absent propeller.

Generally, the effective wake or effective advance ratio of the propeller is secured experimentally by using the open water (or uniform inflow) characterization curves by test of the propeller alone. These curves are entered with the thrust and torque obtained from the same propeller when propelling the model, yielding two different effective advance ratios which are generally averaged. This value is then compared to the volumetric mean from the wake survey. The difference is taken as giving the effective increment in the inflow due to the alterations of the hull wake and propeller loading distribution by the action of the propeller in the radially non-uniform flow. Thus by this procedure the naval architect is given a weighted average inflow in the 'behind condition' whereas the actual effective flow varies quite sharply from root to blade tip.

Although it is known empirically that the efficiency of ship screws is quite insensitive to the radial distribution of pitch, the inception and extent of cavitation is very sensitive to the alignment of the sectional pitches with the flow deviations at various radii. Hence it is important to know the variation in the effective inflow from hub to blade tip for the propeller operating in the spatially non-uniform flow from the hull.

One mechanism through which the propeller alters the flow from that given by the nominal wake measurements is via convection, contraction and stretching of the vorticity inherent in the hull wake.

Among the earliest theoretical studies of the interaction of the flow fields induced by bodies with incident flows having vorticity or shears are those of Tsien (1943) and von Kármán & Tsien (1945). Tsien (1943) treated symmetrical Joukowsky sections in vertically sheared flow and von Kármán & Tsien (1945) dealt with high-aspect-ratio wings in vertical and transverse shears without numerical evaluations. A number of seminal investigations were made by Lighthill (1956, 1957a, b) in part to determine from fundamental fluid mechanics the displacing of the streamlines about a sphere in regard to errors incurred in the use of pitot tubes in shear flows. The effects of shear had been observed experimentally by Young & Maas (1936) and addressed by an approximate theory by Hall (1956). Many studies of sections and wings in shear flows are reported in the aerodynamic literature amongst which are those of Weissinger (1964, 1965a, b) who treated airfoils in linearly and exponentially varying shear.

My involvement in effects of radial shear flow on propellers began in the latter 1970's when the late Dr. W. Cummins, Head of the Hydromechanics Laboratory, David Taylor Model Basin, asked my opinion regarding the disparity of predicted and observed propeller revolutions on submarines. I suggested that the current propeller theories and model test procedures did not account for the effects of the radial shear in the hull wake which also varies with Reynolds number from model to ship. Thereafter we were awarded small contracts to study this aspect which resulted in reports and publications by Dr. T.R. Goodman, my Senior Research Scientist at Davidson Laboratory, Stevens Institute of Technology (Goodman 1979). Further experimental and theoretical correlations were produced by Goodman & Breslin (1982) (under support of the U.S. Maritime Administration) which showed good agreement with laser measurements (made for us in the M.I.T. water tunnel) when a partial allowance was made for non-linear effects.

Other studies of ship propellers in shear flow include those of Cox (1968), Huang *et al.* (1976), Dyne (1982), Falčao de Campos *et al.* (1981), van Gent (1986), Stern *et al.* (1990) and Huang & Groves (1980) (this paper is referred to as *HG* below). All of these studies involve numerical procedures to account for non-linear effects attending deformation of the nominal wake stream surfaces and disc loading.

Here I attempt to provide, in distinction to the completely numerical methods of *HG* and Huang *et al.* (1976), an analytical theory, i.e., one in which the induced components due to shear coupling are given by quadratures which, of course, require computer-effected evaluations. This simplification results from the use of linearized theory and the discovery of a class of shear flows which embraces those from bodies of revolution, appended submarines, and surface ships as determined from wake measurements abaft large models.

9.2   Analysis

9.2.1   *Assumptions*

A basic parameter in the theory is the *propeller-disc loading C_T*, defined by

$$C_T = \frac{T}{\frac{1}{2}\rho U_\infty^2 \pi b^2},\tag{9.1}$$

where T is the specified thrust, ρ is the fluid mass density, b is the propeller radius and U_∞ is the model (or ship) speed (as applies). The following reasonable assumptions are made:

1. The fluid is incompressible and inviscid.
2. The propeller is lightly loaded, i.e., the disc loading C_T satisfies

   $$0 < C_T < 0.6;$$

   this restriction will be relaxed later.
3. The incident flow is axisymmetric about the shaft; it is sheared in the region $h < r \leq s$ and uniform in $s \leq r < \infty$, where h is the propeller hub radius and s is the radial extent of the shear. This flow is assumed to be *independent of the axial coordinate x*.
4. The propeller is representable by an actuator disc of neglible thickness through which fluid flows, and across which there is a distribution of pressure jumps, $\Delta p(r)$ whose integral is T.
5. The nominal wakes are those for single screw hulls without flow separation.
6. The incident flow, $-U(r) \neq 0$ in the entire region, $h \leq r < \infty$.

9.2.2   *Euler and continuity equations*

For an inviscid, incompressible fluid in which the axial onset flow is of the form

$$U(r,\gamma) = \begin{cases} -\displaystyle\sum_{n=-\infty}^{\infty} U_n(r)e^{in\gamma}, & h \leq r \leq s \\[2mm] -U_\infty, & s \leq r < \infty, \end{cases}$$

the Euler equations for steady motion, in cylindrical coordinates (r, γ, x), as displayed in figure 9.1, are

$$(u' - U)(u' - U)_x + \frac{v'}{r}(u' - U)_\gamma + w'(u' - U)_r + \frac{p'_x}{\rho} = -\frac{F_x}{\rho} \quad (9.2)$$

$$(u' - U)v'_x + \frac{v'}{r}v'_\gamma + w'v'_r + \frac{p'_\gamma}{\rho r} = -\frac{F_\gamma}{\rho} \quad (9.3)$$

$$(u' - U)w'_x + \frac{v'}{r}w'_\gamma + w'w'_r + \frac{p'_r}{\rho} = 0 \quad (9.4)$$

where the applied forces on an element of the actuator disc are taken to be

$$F_x = \frac{1}{r}\Delta p(r', \gamma')\delta(x)\delta(r - r')\delta(\gamma - \gamma')dS' \cos\beta',$$

$$F_\gamma = \frac{1}{r}\Delta p(r', \gamma')\delta(x)\delta(r - r')\delta(\gamma - \gamma')dS' \sin\beta',$$

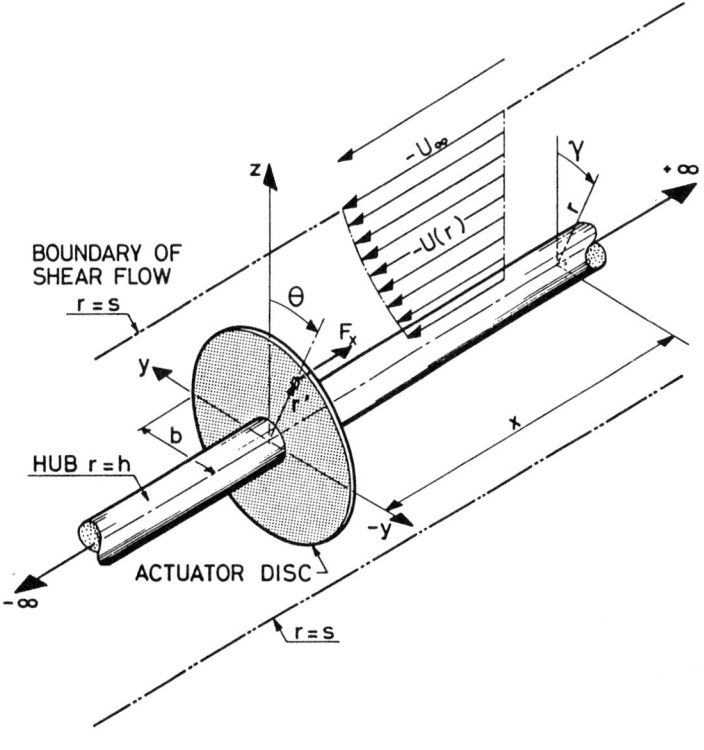

Figure 9.1: Actuator disc representation of a propeller in an axisymmetric shear flow $-U(r)$

u', v', w', p' are the perturbed velocity components and pressure due to these elementary forces; $\delta(\ )$ is the Dirac delta function, β' is the fluid pitch angle and dS' is an element of area of the disc. The primes on the induced quantities will be removed later by integration over the disc.

To these equations we must append the continuity equation

$$u'_x + \frac{1}{r}v'_\gamma + w'_r + \frac{w'}{r} = 0 \tag{9.5}$$

9.2.3   Axisymmetric linearized equations

From now on, we consider the axisymmetric case, in which there is no dependence on γ, and so we take the zero harmonics ($n = 0$) throughout. Using a subscript zero to denote the corresponding solution and limiting to small disc loadings, so that $u'_0/U_0 \ll 1$ and $w'_0/U_0 \ll 1$, we retain

$$-U_0 u'_{0x} - w'_0 U'_0(r) + \frac{p'_{0x}}{\rho} = \frac{-\Delta p(r')}{2\pi\rho r}\delta(x)\delta(r - r')dS' \tag{9.6}$$

$$-U_0 w'_{0x} + \frac{p'_{0r}}{\rho} = 0 \tag{9.7}$$

where $\beta' = 0$ has been taken, $dS' = 2\pi r'dr'$ and

$$U'_0(r) = \frac{dU_0}{dr} \tag{9.8}$$

(Note is taken that the linearization becomes tenuous for small r as $U_0(r)$ become small as $r \to 0$).

Eliminating the velocity components with the aid of the continuity equation (9.5) and dropping the subscript zero for ease of transcription gives

$$\nabla^2 p' - 2\frac{U'(r)}{U(r)}p'_r = -\frac{\Delta p(r')}{2\pi r}\delta'(x)\delta(r - r')dS' \tag{9.9}$$

where

$$\nabla^2 p' = p'_{xx} + p'_{rr} + \frac{1}{r}p'_r \quad \text{and} \quad \delta'(x) = \frac{d\delta}{dx}(x).$$

The following transformation (by T.R. Goodman),

$$\psi = V(r)p'(x,r) \quad \text{with} \quad V = \frac{1}{U(r)}, \tag{9.10}$$

exactly converts (9.9) into

$$\nabla^2\psi - \left(\frac{\nabla_1^2 V}{V}\right)\psi = \frac{-\Delta p}{2\pi r}V\delta'(x)\delta(r - r')dS' \tag{9.11}$$

where

$$\nabla_1^2 = \frac{d^2}{dr^2} + \frac{1}{r}\frac{d}{dr}.$$

Equation (9.11) is an inhomogeneous, modified Helmholtz equation which can be changed to an integral equation, involving numerical inversion for arbitrary $V(r)$.

9.2.4   Solution for a class of shear flows

I assume the existence of a class of shear flows defined by

$$\frac{\nabla_1^2 V}{V} = \frac{1}{V}\left(\frac{d^2}{dr^2} + \frac{1}{r}\frac{d}{dr}\right)V = k^2 = \text{a constant.} \tag{9.12}$$

This is a Bessel equation whose general solution is,

$$V = AI_0(kr) + BK_0(kr) = 1/U(r), \tag{9.13}$$

where I_0 and K_0 are modified Bessel functions of the first and second kind of order zero, respectively.

Equation (9.13) contains three unknown parameters, A, B, and k. It is now necessary to show that (9.13) can fit measured $U(r)$ over the entire range of data for which $U'(r)$ is non-zero.

It has been found via computer-effected calculations by Dr. E.D. Park and Mr. S.S. Lee in South Korea that the *four* parameters, A, B, k and s (the effective radius of the shear flow) can be determined to fit very closely measured values of $U(r)$ from wake surveys on large models of bodies of revolution, appended submarines as well as surface ships!

Four conditions are imposed:

$$
\begin{aligned}
AI_0(kr_1) + BK_0(kr_1) &= 1/U(r_1); & \text{(measured) } r_1 < s; \\
AI_0(kr_2) + BK_0(kr_2) &= 1/U(r_2); & \text{(measured) } r_2 < s; \\
AI_0(ks) + BK_0(ks) &= 1/U_\infty(s); & \text{(measured) } r = s \\
AI_1(ks) - BK_1(ks) &= -U'(s)/U^2(s) = 0
\end{aligned}
$$

Computed values of A, B, k and s are listed for three nominal wakes in Table 9.1. The root-mean-square deviation also listed shows this measure of the error of the fit to be about 1/2 percent! Graphical comparisons of measured nominal wakes with these computed fits of (9.13) are shown in figures 9.2, 9.3 and 9.4. Excellent agreement is displayed over the entire range of available data which admittedly does not extend to the region where $U(r)/U_\infty$ approaches unity. It is realized that the shear flow vanishes asymptotically but forcing our substitute class of shear flow to a ratio of unity at a finite radius appears admissible for my purpose because $U'(r)$ is rapidly vanishing at large radii.

Equation (9.11) for this class of shear flows reduces to

Wake	k	A	B	s/b	RMS%
1	0.8363	0.4346	1.3375	1.7260	0.530
C	0.6307	0.5083	0.9176	1.9583	0.546
D	0.6630	0.4698	1.1159	2.0218	0.530

Table 9.1: Computed parameters for the approximation of three nominal wakes (1, C and D) by (9.13).

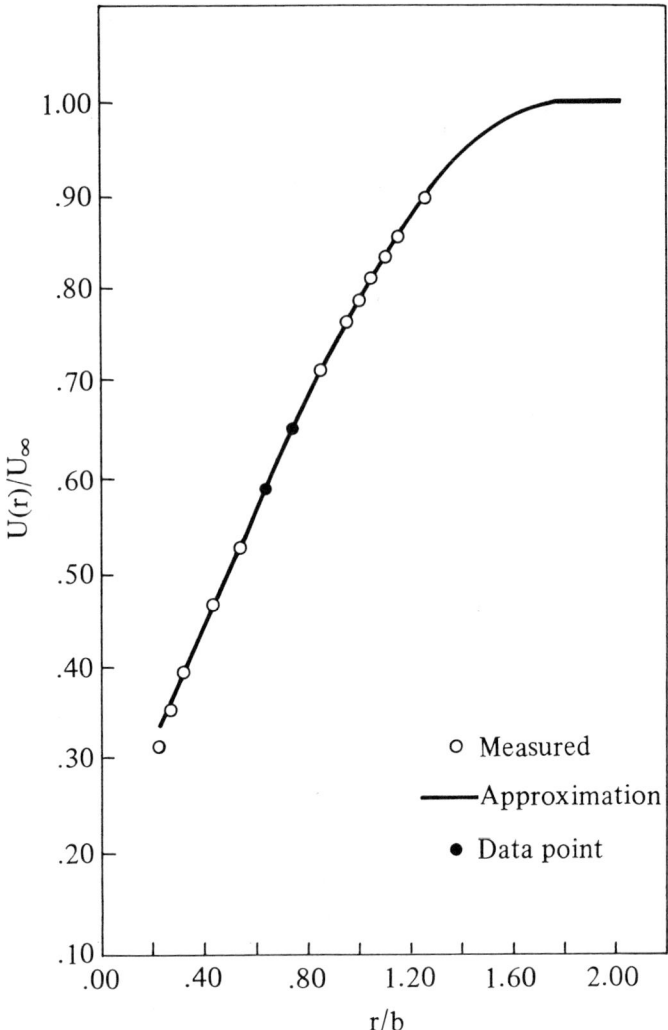

Figure 9.2: Comparison of measured nominal wake 1 and fit of equation (9.13) to measured values.

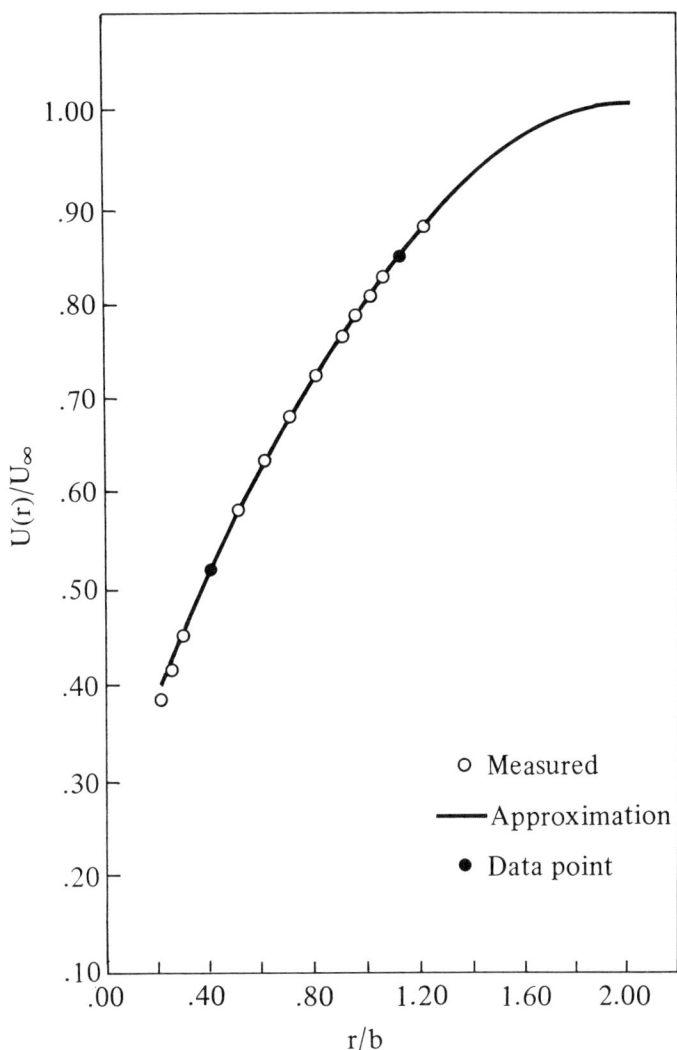

Figure 9.3: Comparison of measured nominal wake C and fit of equation (9.13) to measured values.

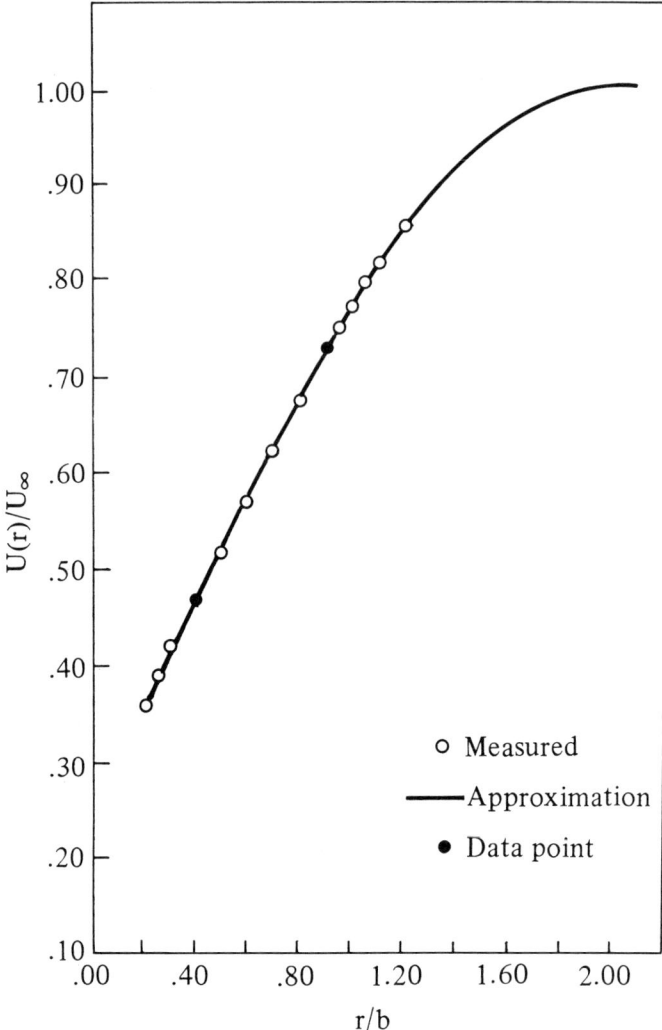

Figure 9.4: Comparison of measured nominal wake D and fit of equation (9.13) to measured values.

$$\left(\frac{\partial^2}{\partial x^2} + \frac{\partial^2}{\partial r^2} + \frac{1}{r}\frac{\partial}{\partial r} - k^2\right)\psi = -\frac{V(r)}{2\pi r}\Delta p\,\delta'(x)\delta(r-r')dS' \quad (9.14)$$

for $h \le r \le s$ and, for the region outboard of the shear,

$$\left(\frac{\partial^2}{\partial x^2} + \frac{\partial^2}{\partial r^2} + \frac{1}{r}\frac{\partial}{\partial r}\right)\psi_0 = -\frac{V(r)}{2\pi r}\Delta p\,\delta'(x)\delta(r-r')dS'$$
$$= 0 \quad (9.15)$$

for $s \le r < \infty$, since $r' < s$. Application of the x-wise Fourier transform reduces (9.14) to an inhomogeneous Bessel equation and (9.15) to a homogeneous Bessel equation:

$$\widehat{\psi}_{rr} + \frac{1}{r}\widehat{\psi}_r - (\xi^2 + k^2)\widehat{\psi} = \frac{-i}{2\pi r}\xi V(r)\Delta p\delta(r-r')dS' \quad (9.16)$$

for $h < r < s$ and

$$\widehat{\psi}_{0rr} + \frac{1}{r}\widehat{\psi}_{0r} - \xi^2\widehat{\psi}_0 = 0 \quad (9.17)$$

for $s < r < \infty$, where

$$\widehat{\psi}(r;\xi) = \int_{-\infty}^{\infty} \psi(x,r)e^{-i\xi x}\,dx.$$

Solutions are sought which meet the following boundary conditions.

- For the outer region, $s \le r < \infty$:
$$p(x,r) \to 0 \text{ and } \psi_0 \to 0 \text{ as } r \to \infty. \quad (9.18)$$

- On the shear-flow boundary, $r = s$, there are two conditions:
  1. Continuity of pressure over all x:
  $$\widehat{\psi}(s) = \widehat{\psi}_0(s). \quad (9.19)$$
  2. Continuity of radial velocity which implies continuity of p_r:
  $$\widehat{\psi}_r = \widehat{\psi}_{0r}. \quad (9.20)$$

- On the hub surface, $r = h$, $w'(x,h) = 0$, all x:
$$U'(h)\widehat{\psi}(h) + U(h)\widehat{\psi}_r(h) = 0. \quad (9.21)$$

The solution $\widehat{\psi}_0$ which satisfies (9.17) and condition (9.18) is

$$\widehat{\psi}_0(r;\xi) = \tilde{a}K_0(\xi r). \quad (9.22)$$

The solution of (9.16), by the method of variation of parameters, is

$$\widehat{\psi} = \tilde{b}I_0(\lambda r) + \tilde{c}K_0(\lambda r)$$
$$+ i\xi\left\{\int_0^r \frac{\delta(r''-r')}{r''\lambda W(\lambda r'')}K_0 r'')V(r'')I_0(\lambda r)\,dr'' \right. \quad (9.23)$$
$$\left. - \int_0^r \frac{\delta(r''-r')}{r''\lambda W(\lambda r'')}I_0(\lambda r'')V(r'')K_0(\lambda r)\,dr''\right\}\Delta p(r')r'dr'$$

where $\lambda = \sqrt{\xi^2 + k^2}$ and $W = -1/(\lambda r'')$ is the Wronskian. For $h \le r \le r'$ the r''-integrals do not contribute leaving

$$\hat{\psi}(r;\xi) = \tilde{b}I_0(\lambda r) + \tilde{c}K_0(\lambda r). \tag{9.24}$$

For $r' \le r \le s$, we obtain

$$\begin{aligned}
\hat{\psi}(r;\xi) &= \tilde{b}I_0(\lambda r) + \tilde{c}K_0(\lambda r) - i\xi\{K_0(\lambda r')I_0(\lambda r) \\
&\quad - I_0(\lambda r')K_0(\lambda r)\}V(r')\Delta p(r')r'dr'.
\end{aligned} \tag{9.25}$$

The three unknown coefficients $\tilde{a}$, $\tilde{b}$ and $\tilde{c}$ are determined by enforcing the two conditions on $r = s$, (9.19) and (9.20) and the single requirement on the hub, $r = h$ given by (9.21). These details are relegated to the Appendix.

Application of the inverse Fourier transform,

$$\psi(x,r) = \frac{1}{2\pi}\int_{-\infty}^{\infty}\hat{\psi}(r;\xi)e^{i\xi x}\,d\xi \tag{9.26}$$

and integration over the disc employing the appropriate parts of $\hat{\psi}$ gives the solution for the pressure in the form:

$$\begin{aligned}
p(x,r) &= \frac{U(r)}{2\pi}\int_h^b D\,\Delta p\,r'\,dr' \\
&\quad + \frac{U(r)}{2\pi}\int_h^r E\,\Delta p\,V(r')r'\,dr'
\end{aligned} \tag{9.27}$$

for $h \le r \le s$, where

$$\begin{aligned}
D &= \int_{-\infty}^{\infty}\{b'(\xi,k)I_0(\lambda r) + c'(\xi,k)K_0(\lambda r)\}e^{i\xi x}\,d\xi, \\
E &= \int_{-\infty}^{\infty}i\xi\{I_0(\lambda r')K_0(\lambda r) - K_0(\lambda r')I_0(\lambda r)\}e^{i\xi x}\,d\xi,
\end{aligned}$$

$\tilde{b} = b'\Delta p(r')r'dr'$ and $\tilde{c} = c'\Delta p(r')r'dr'$.

It can be shown that p exhibits the required behaviours, namely,

$$\lim_{x\to\pm 0}p(x,r) = \mp\Delta p(r)/2 \qquad \text{for } h \le r \le b \text{ and} \tag{9.28}$$

$$\lim_{x\to\pm\infty}p(x,r) = 0. \tag{9.29}$$

We may note that p does not depend explicitly on $U'(r)$ but is a functional of the shear through the shear parameter, k.

9.2.5   The radial component of induced velocity

Integration of the radial equation of motion gives,

$$w(x,r;U',k) = \frac{-1}{\rho U(r)}\int_x^{\infty}p_r(x',r)\,dx'. \tag{9.30}$$

This is employed to obtain the axial component after integration over the actuator disc.

9.2.6   The axial component of induced velocity
Integration of the axial equation of motion, using (9.30), yields

$$u(x,r;U',k) = \frac{p(x,r;k)}{\rho U(r)} + \mathcal{U}(x,r)$$
$$- \frac{U'}{\rho U^2(r)} \int_x^\infty \int_{x''}^\infty p_r(x',r)\,dx'\,dx'', \qquad (9.31)$$

where

$$\mathcal{U} = \begin{cases} 0, & x > 0, \\ \frac{\Delta p(r)}{\rho U(r)}, & x < 0, \quad h \le r \le b. \end{cases}$$

As p varies as $U(r)F(x,r)$ we see that the third term yields terms proportional to U' and U'^2. The expression for u can be shown to be continuous through the disc as required physically.

In the absence or neglect of shear, $U' = 0$, $k = 0$, the axial component is,

$$u(x,r;0,0) = \frac{p(x,r;0)}{\rho U(r)} - \mathcal{U}(x,r). \qquad (9.32)$$

Then the increment due to shear is, for equal loadings,

$$\Delta u_s(x,r) = u(x,r;U',k) - u(x,r;0,0)$$
$$= \frac{p(x,r;k) - p(x,r;0)}{\rho U(r)}$$
$$- \frac{U'(r)}{\rho U^2(r)} \int_x^\infty \int_{x''}^\infty p_r(x',r)\,dx'\,dx'' \qquad (9.33)$$

The addition of Δu_s to the nominal axial wakes gives an effective wake. It can be shown that on the disc, $x = 0$,

$$\lim_{x \to 0} p(x,r;k) = \mp \Delta p/2 = \lim_{x \to 0} p(x,r;0),$$

so that the effective wake on the disc in linear theory is

$$\frac{U_e}{U_\infty} = -\frac{U(r)}{U_\infty} - \frac{U'(r)}{\rho U_\infty U^2(r)} \int_0^\infty \int_{x''}^\infty p_r(x',r)\,dx'\,dx''. \qquad (9.34)$$

Effecting the integrations with respect to x' and x'' gives

$$\frac{U_e}{U_\infty} = -\frac{U(r)}{U_\infty} - \frac{U'(r)}{2U_\infty} \int_h^b \mathcal{F}_1(r,r')\frac{\Delta p}{\rho U_\infty^2} r'\,dr'$$
$$+ \frac{U'(r)}{2U_\infty} \int_h^r \mathcal{F}_2(r,r')\frac{\Delta p}{\rho U_\infty^2} r'\,dr' \qquad (9.35)$$

where the functions $\mathcal{F}_1$ and $\mathcal{F}_2$ are known explicitly and are given in the Appendix. It may be noted that $U_e(h) = -U(h)$ as a consequence of the boundary condition on the hub.

9.3   Numerical evaluations

We take a class of radial loadings which represents practical thrust gradients in the form

$$dT/dr = c(r - h)\sqrt{b - r}$$

where c is a constant, h is the hub radius and b is the radius of the disc. Alternatively, we have

$$dT/dr = 2\pi r \Delta p(r).$$

We determine c by integration over the disc to obtain,

$$c = \frac{T}{\int_h^b (r - h)\sqrt{b - r}\, dr}.$$

Using the definition of propeller-disc loading, (9.1),

$$T = \frac{1}{2}\rho \pi b^2 U_\infty^2 C_T,$$

we have

$$\frac{\Delta p}{\rho U_\infty^2} = \frac{15}{16} C_T \frac{(r - h)\sqrt{1 - r}}{r(1 - h)^{5/2}}. \tag{9.36}$$

Evaluations have been made by Dr. E.D. Park and S.S. Lee at the Chinhae Machine Depot, South Korea for three nominal wakes and seven values of C_T as employed by Huang & Groves (1980; HG) at David Taylor Research Center. The numerical results given here were obtained *without enforcing the hub condition.*

Comparisons on the disc with HG are displayed in the figures 9.5–9.8 for representative conditions.

9.4   Partly non-linear theory

The foregoing comparisons with HG show disagreements with their curves at the inner radii which departures increase sharply with increasing values of propeller-disc loading C_T. These departures are caused primarily by the inherent non-linearity of the axial velocity induced by an actuator disc *in the absence of shear!*

By retaining the non-linear term $u'u'_x$ in (9.2), we can readily secure the following increment due to shear at the disc:

$$\frac{(\Delta u_s)_{\text{pnl}}}{U_\infty} = \frac{U(r)}{U_\infty}\left\{\sqrt{1 + \frac{\Delta p}{\rho U^2}} - \sqrt{1 + \frac{\Delta p}{\rho U^2} + \frac{2u_s}{U}}\right\} \tag{9.37}$$

where

$$u_s = \frac{U'(r)}{\rho U^2(r)} \int_0^\infty \int_{x''}^\infty p_r(x', r)\, dx'\, dx''$$

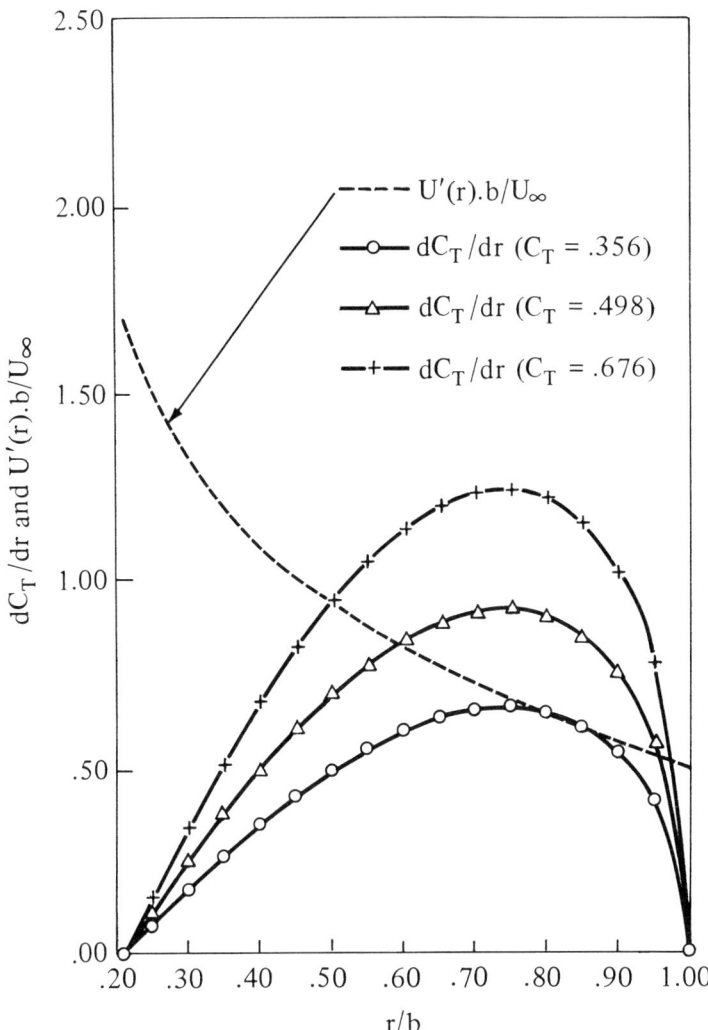

Figure 9.5: Thrust density distributions and radial variation of shear parameter $U'(r)b/U_\infty$ for Huang's Wake C.

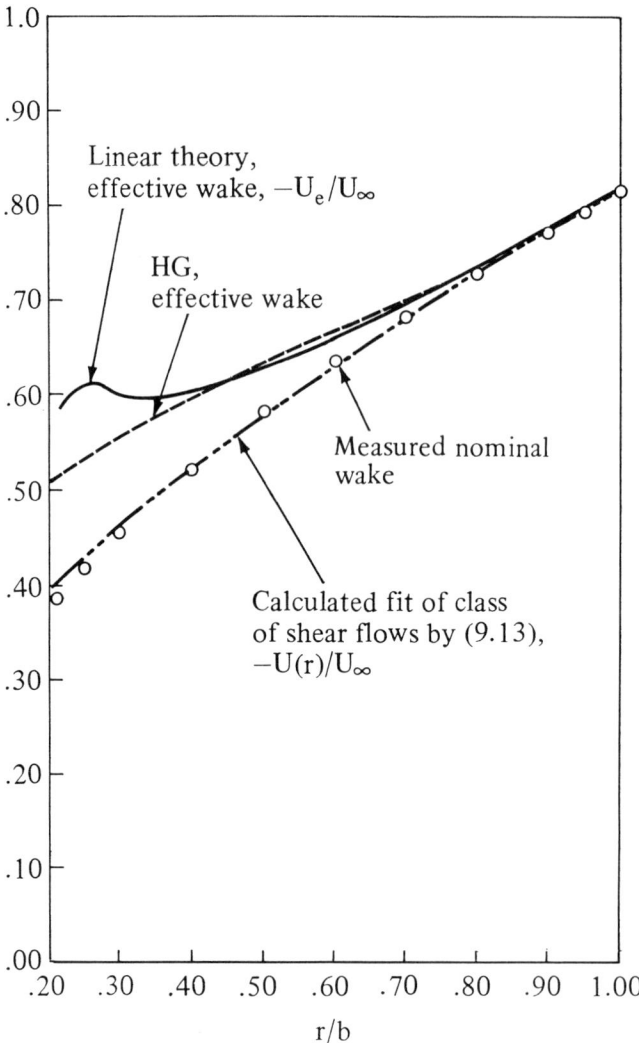

Figure 9.6: Nominal and effective wakes within the propeller disc for Huang's wake C and $C_T = 0.356$

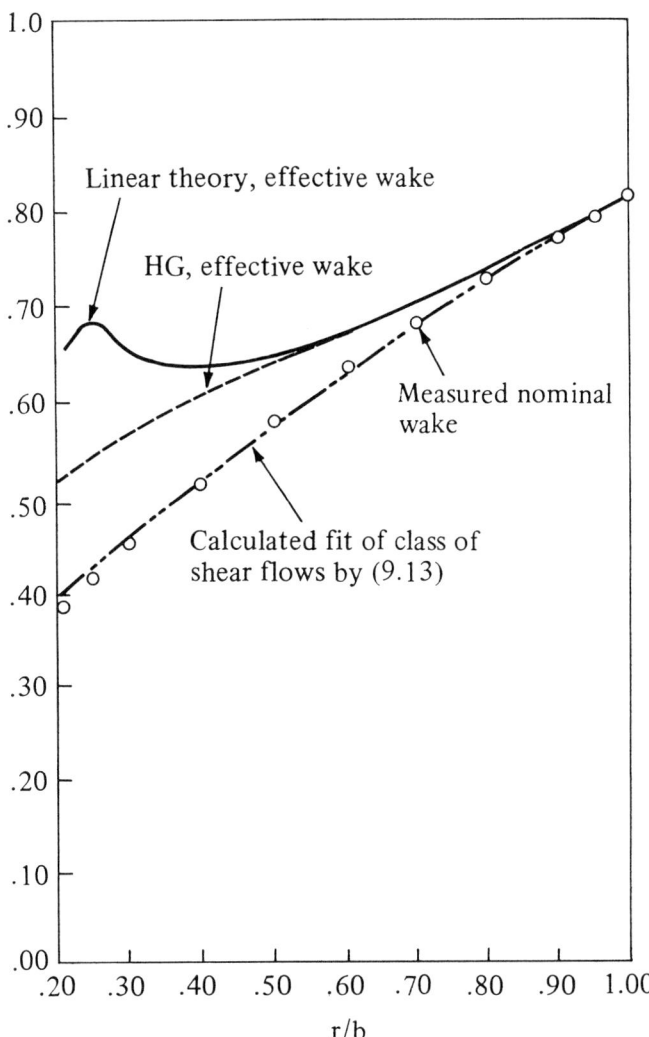

Figure 9.7: Nominal and effective wakes within the propeller disc for Huang's wake C and $C_T = 0.498$

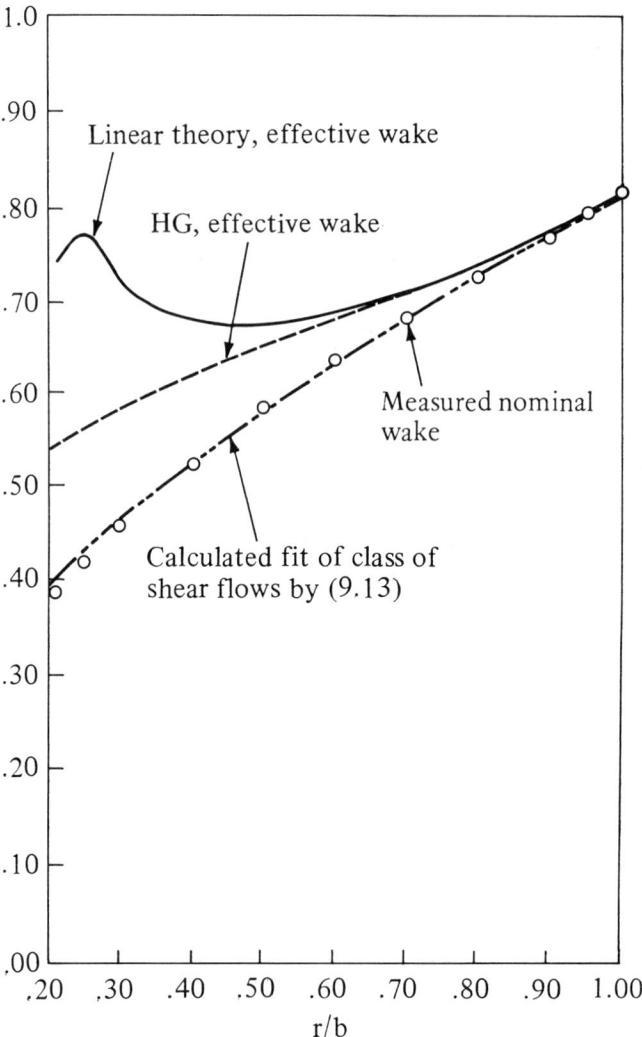

Figure 9.8: Nominal and effective wakes within the propeller disc for Huang's wake C and $C_T = 0.676$

is the induced component generated by shear coupling given by the *linearized* equations.

Non-linear terms in the radial equation are believed to be weaker than that retained in the axial equation. Admittedly, this accounting for the non-linear induction of the actuator disc is inconsistent. It may be justified in part from the use of such an inconsistent theory in texts on theoretical naval architecture where applications for moderate disc loadings show good agreement with measurements in the absence of shear. Surely for large disc loading coefficients a completely non-linear theory must be applied.

This modified theory then gives the following prescription for an effective wake:

$$\frac{U_e}{U_\infty} = \frac{-U(r)}{U_\infty} + \left(\frac{\Delta u_s}{U_\infty}(0, r; k) \right)_{\text{pnl}}. \tag{9.38}$$

Comparisons of curves produced from (9.38) with results of HG as displayed in figures 9.9 and 9.10 for two disc loadings show excellent agreement over almost the entire range.

Had the effect of the hub been included, then the curve of effective wake would join that of the nominal wake at $r = h = 0.2$. This is because on the hub $p_r(x', h) = 0$ identically and $\Delta p(h) = 0$ by construction. (Unfortunately, calculations including the hub boundary condition have yet to be completed by Dr. T.G. McKee at Davidson Laboratory, Stevens Institute of Technology, both on and forward of the disc. The latter calculated values are essential for comparison with measured components.)

9.5   Comparison of weighted-average wake parameters

Weighted-average nominal and effective wakes (as used by naval architects and ship-model basins) as defined below have been calculated by Park and Lee for three wakes treated in HG. The values computed from HG are compared with those derived from the present calculations in Table 9.2. This table also gives the effective values obtained from each set at the 0.7 radius, values of the advance ratio J_V and values of $1 - W_T$, where W_T is the Taylor wake fraction. Here,

$$J_V = \frac{U_\infty}{ND},$$

where the model shaft rotates N times per second and carries a propeller of diameter D, and W_T is determined experimentally using a self-propulsion test and the open-water propeller characteristics.

The volumetric means of the nominal and effective wakes are defined as follows:

$$\left(\frac{U}{U_\infty} \right)_m = \frac{2\pi \int_h^b (U(r)/U_\infty) r \, dr}{\pi(b^2 - h^2)}, \tag{9.39}$$

$$\left(\frac{U_e}{U_\infty} \right)_m = \frac{2\pi \int_h^b (U_e/U_\infty) r \, dr}{\pi(b^2 - h^2)}. \tag{9.40}$$

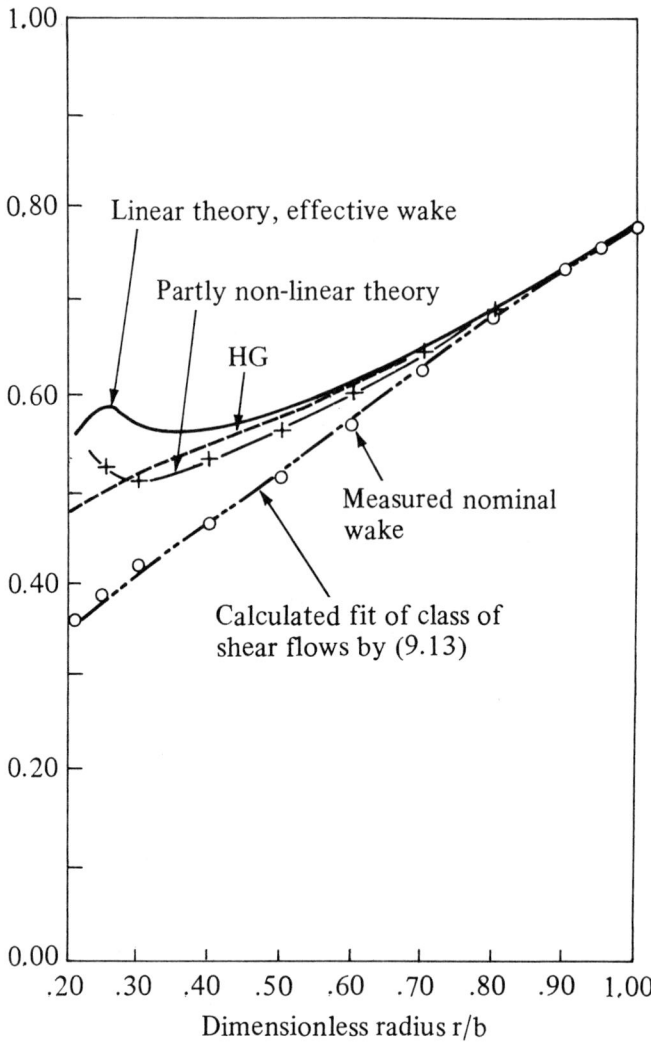

Figure 9.9: Nominal and effective axial wakes within the propeller disc, for Huang's wake D and disc loading $C_T = 0.360$.

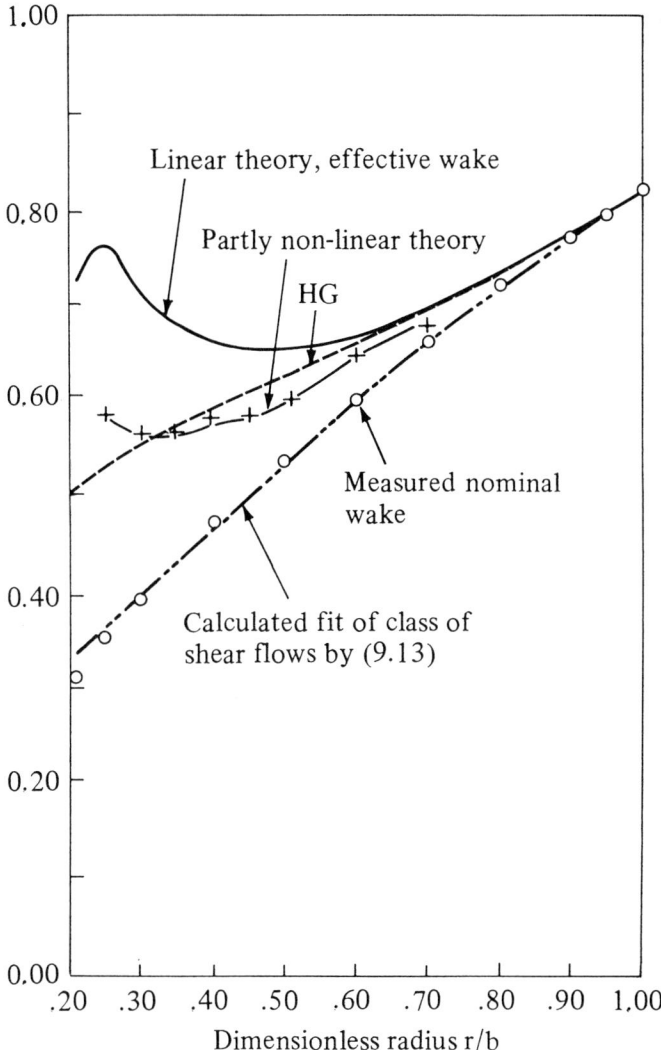

Figure 9.10: Nominal and effective axial wakes within the propeller disc, for Huang's wake 1 and $C_T = 0.654$.

Here it is evident that the calculated values from the present linearized theory show remarkable agreement with both the determinations of HG as well as those derived from model measurements! The discrepancies in the inner radii are, of course, muted by the weighting factor r in the definition of the mean values. It is also remarkable that the values of effective wake at the 0.7 radius are very close to the volumetric means from the experiment. Most assuredly the results from the present partly non-linear theory would be even closer to both HG and those from measurements.

9.6   Concluding remarks

The foregoing analysis and evaluations demonstrate that useful predictions can be made of effective wake at light and moderate disc loadings by means of this theory involving formulæ in terms of Bessel functions for which computer subroutines exist. The theory does not extend to heavily loaded propellers nor can it be applied to cases where the hull boundary layer separates forward of the propeller. It also requires that the nominal wake be known in advance. It is also necessary to emphasize that the propeller thrust and the radial distribution of loading are functionals of the effective inflow. This requires interative calculations through the use of a propeller program (lifting surface or at least a lifting line procedure) which gives the thrust and loading distribution for a known velocity input. Further comparisons should be made with the results of Stern *et al.* (1990) who treat the development and interaction with the propeller of the turbulent flow over the afterbody of a body of revolution. They also point out that the effective inflow in HG is considerably larger than theirs in the inner radii and indeed becomes equal to the nominal inflow very near to the hull surface. Had the present theory been evaluated with the hub boundary condition invoked it would show juncture of effective and nominal inflows at the hub. Such correlations with the hub condition

	$\left(\frac{U}{U_\infty}\right)_m$	J_V	C_T	$1 - W_T$	Huang & Groves		Present	
					$\left(\frac{U_e}{U_\infty}\right)_m$	$\left(\frac{U_e}{U_\infty}\right)_{0.7}$	$\left(\frac{U_e}{U_\infty}\right)_m$	$\left(\frac{U_e}{U_\infty}\right)_{0.7}$
1	0.640	1.250	0.370	0.685	0.676	0.672	0.681	0.671
		1.070	0.654	0.695	0.689	0.683	0.713	0.688
C	0.665	1.268	0.356	0.694	0.694	0.694	0.698	0.693
		1.157	0.498	0.704	0.699	0.698	0.711	0.699
		1.066	0.676	0.706	0.704	0.704	0.727	0.707
D	0.615	1.290	0.360	0.667	0.642	0.642	0.655	0.646
		1.174	0.500	0.681	0.655	0.649	0.700	0.653
		1.054	0.706	0.685	0.661	0.654	0.691	0.664

Table 9.2: Comparison of computed effective wake parameters with Huang & Groves (1980) and values deduced from propeller measurements at various propulsion conditions in three wakes.

as well as iterative calculations of the loading will be made in the near future.

9.7 Acknowledgements

Support of this work provided by the Office of Naval Research, Department of the United States Navy, under the Special Focus Program is gratefully acknowledged. Dr. E.D. Park and Mr. S.S. Lee of the Chinhae Machine Depot, Chinhae, South Korea have provided the numerical evaluations of this theory for which I am most thankful. Professor J.N. Newman is thanked for providing Weissinger's reports of which I was unaware. I am indebted to Dr. Paul Martin for his careful and constructive editing of this paper.

9.8 References

Cox, B.D. 1968 Vortex ring solutions of axisymmetric propeller flow problems. Massachusetts Inst. of Technology, Dept. of Naval Arch. & Marine Engng., Rpt. 68-13.

Dyne, G. 1982 Non-linear interaction between an actuator disc, an axisymmetric afterbody and a simplified wake. Swedish Maritime Res. Ctr., SSPA Rpt. 2363-3.

Falcão de Campos, J.A.C., & van Gent, W. 1981 Effective wake of an open propeller in axisymmetric shear flow. Netherlands Ship Model Basin Rpt. 50030-SR.

Goodman, T.R. 1979 Momentum theory of a propeller in a shear flow. *J. Ship Res.* **23**, 242–252.

Goodman, T.R., & Breslin, J.P. 1982 Theoretical and experimental induction generated by a propeller in an axisymmetric shear flow. Stevens Inst. of Technology, Dept. of Ocean Engng., Rpt. SIT-OE-82-2.

Hall, I.M. 1956 The displacement effect of a sphere in a two-dimensional shear flow. *J. Fluid Mech.* **1**, 142–162.

Huang, T.T., Wang, H.T., Santelli, N., & Groves, N.C. 1976 Propeller/stern boundary interaction on axisymmetric bodies: theory and experiment. D.W. Taylor Naval Ship R&D Center, Rpt. 76-0113.

Huang, T.T., & Groves, N.C. 1980 Effective wake: theory and experiment. *Proc. 13th Symposium on Naval Hydrodynamics*, Tokyo, 651–669.

Lighthill, M.J. 1956 Drift. *J. Fluid Mech.* **1**, 31–52.

Lighthill, M.J. 1957a Contributions to the theory of the Pitot-tube displacement effect. *J. Fluid Mech.* **2**, 493–512.

Lighthill, M.J. 1957b The fundamental solution for small steady three-dimensional disturbances to a two-dimensional parallel shear flow. *J. Fluid Mech.* **3**, 113–144.

Stern, F., Toda, Y., & Kim, H.T. 1990 Computations of viscous flow around propeller-body configurations–Part 1: Iowa axisymmetric body. *J. Ship Res.*, submitted.

Tsien, H.S. 1943 Symmetrical Joukowsky airfoils in shear flow. *Quart. Appl. Math.* **1**, 130–148.

van Gent, W. 1986 A model of propeller-ship wake interaction. Maritime Res. Inst. Netherlands (MARIN), Rpt. P50625-1-RF.

von Kármán, T., & Tsien, H.S. 1945 Lifting-line theory for a wing in non-uniform flow. *Quart. Appl. Math.* **3**, 1–11.

Weissinger, J. 1964 Non-uniform steady flow of an ideal fluid past airfoils. Part I: Some exact solutions for two-dimensional non-uniform flow of exponential type. Math. Res. Ctr., University of Wisconsin, Rpt. 515.

Weissinger, J. 1965a Non-uniform steady flow of an ideal fluid past airfoils. Part II: Approximate theory for thin airfoils in exponential flow. Math. Res. Ctr., University of Wisconsin, Rpt. 536.

Weissinger, J. 1965b Non-uniform steady flow of an ideal fluid past airfoils. Part III: Numerical results. Math. Res. Ctr., University of Wisconsin, Rpt. 571.

Young, A.D., & Maas, J.N. 1936 The behaviour of a Pitot tube in a transverse total-pressure gradient. Aeronautical Research Committee, London, *Rpts. & Memoranda* No. 1770.

Appendix. Determination of the coefficients $\tilde{a}, \tilde{b}$ and $\tilde{c}$

As these coefficients are all proportional to the pressure jump Δp and to the area element $r'dr'$ we replace them with

$$\tilde{a} = a'\Delta p\, r'dr', \quad \tilde{b} = b'\Delta p\, r'dr' \quad \text{and} \quad \tilde{c} = c'\Delta p\, r'dr'.$$

Enforcing the boundary conditions specified by (9.19), (9.20) and (9.21) upon (9.22), (9.24) and (9.25) gives the following simultaneous, linear, algebraic equations for a', b' and c':

$$\alpha_{11}a' + \alpha_{12}b' + \alpha_{13}c' = F$$
$$\alpha_{21}a' + \alpha_{22}b' + \alpha_{23}c' = \lambda F'$$
$$0 + \alpha_{32}b' + \alpha_{33}c' = 0$$

where

$$\alpha_{11} = K_0(\xi s), \quad \alpha_{12} = -I_0(\lambda s), \quad \alpha_{13} = -K_0(\lambda s),$$

$$\alpha_{21} = |\xi|K_0'(\xi s), \quad \alpha_{22} = -\lambda I_0'(\lambda s), \quad \alpha_{23} = -\lambda K_0'(\lambda s),$$

$$\alpha_{32} = (UI_0)_h', \quad \alpha_{33} = (UK_0)_h',$$

$$F = i\xi\{K_0(\lambda r')I_0(\lambda s) - I_0(\lambda r')K_0(\lambda s)\}V(r'),$$

$$(UI_0)_h' = \frac{d}{dr}(U(r)I_0(\lambda r))\,|_{r=h},$$

$$(U K_0)_h' = \frac{d}{dr}(U(r)K_0(\lambda r)) \mid_{r=h},$$

$$F' = F_r \quad \text{at} \quad r = s.$$

For present purposes we need b' and c' for the region of interest occupied by the disc, $h \leq r \leq b < s$. The resulting expressions are:

$$b' = -i\xi(U K_0)_h' V(r')\frac{N_e}{D_e} \quad \text{and} \quad c' = +i\xi(U I_0)_h' V(r')\frac{N_e}{D_e},$$

where

$$
\begin{aligned}
N_e &= \{\lambda K_0(\xi s)K_0'(\lambda s) - |\xi|K_0'(\xi s)K_0(\lambda s)\}I_0(\lambda r') \\
&\quad + \{|\xi|K_0'(\xi s)I_0(\lambda s) - \lambda K_0(\xi s)I_0'(\lambda s)\}K_0(\lambda r') \\
D_e &= \{|\xi|K_0'(\xi s)K_0(\lambda s) - \lambda K_0'(\lambda s)K_0(\xi s)\}(U I_0)_h' \\
&\quad + \{\lambda I_0'(\lambda s)K_0(\xi s) - |\xi|K_0'(\xi s)I_0(\lambda s)\}(U K_0)_h'.
\end{aligned}
$$

It is easy to show that for $h = 0$, $c' = 0$ as required, and that for $h = 0 = k$, the pressure becomes that for an actuator disc in an unbounded uniform stream, as required physically.

The above results are used in (9.34) to yield (9.35), where

$$\mathcal{F}_1 = \frac{V(r')\mathcal{G}(r)\mathcal{H}(r')}{K_1(ks)(U I_0)_h' + I_1(ks)(U K_0)_h'},$$

$$\mathcal{F}_2 = V(r')\left[K_0(kr')(U I_0)_r' - I_0(kr')(U K_0)_r'\right],$$

$$\mathcal{G}(r) = (U I_0)_h'(U K_0)_r' - (U K_0)_h'(U I_0)_r',$$

and

$$\mathcal{H}(r') = K_1(ks)I_0(kr') + I_1(ks)K_0(kr').$$

Note that $\mathcal{G}(h) = 0$.

10

Analytic aspects of slender body theory

E.O. Tuck
University of Adelaide

> *When I started my Ph.D. under Fritz Ursell's supervision in 1960, the application of aerodynamic slender body concepts to computation of the wave resistance of slender ships was my first and best choice among the topics that he suggested. As it happened, my own highly intuitive approach to slender body theory, via matched asymptotic expansions, did not entirely suit Ursell's more rigorous personal style. But even within an intuitive framework, there was much analytical work to be done, and no better teacher than Ursell. Fritz and I worked well together then, and have done so again since, on various analytical problems arising in water wave theory and other topics, as well as in slender body theory. In the present article, I want to revive and illuminate some of the analytic issues that are both central and peripheral to slender body theory, while avoiding matched expansions as far as possible.*

10.1   The central asymptotic result of slender body theory

Slender body theory depends essentially on the following remarkably simple asymptotic formula, valid for an arbitrary given function $m(x)$ in the limit as $r \to 0$:

$$-\frac{1}{4\pi} \int_{-\infty}^{\infty} \frac{m(\xi)\,d\xi}{\sqrt{(x-\xi)^2 + r^2}} = \frac{m(x)}{2\pi} \log r + f(x) + O(r^2 \log r). \tag{10.1}$$

The above result is only 'simple' because we have not yet specified the actual form of the function $f(x)$ that appears on the right. Before making the result look more formidable than it really is by doing so, let us motivate the result in hydrodynamic terms; however, bear in mind that (10.1) is a formal limit that is independently derivable and of mathematical value in its own right, irrespective of applications.

The quantity on the left of (10.1) satisfies Laplace's equation, and is the velocity potential $\phi(x,y,z)$, with $r^2 = y^2 + z^2$, for a distribution of three-dimensional point sources of inviscid incompressible fluid along the x-axis, the source strength per unit length at $(x, y, z) = (\xi, 0, 0)$ being the given function $m(\xi)$. The first term on the right of (10.1) is the velocity potential for a two-dimensional line source in the (y, z) plane at fixed x, whose strength $m(x)$ is constant in that plane and exactly equal to the local strength per unit length of the original three-dimensional sources. The term $f(x)$ is, for fixed x, just an additive 'constant' in the velocity potential, and has no significance from the point of view of such a purely two-dimensional flow.

From this point of view, the result (10.1) is entirely natural — three dimensional source distributions along a line look like two dimensional line sources when we are close to them. The constants $-1/4\pi$ and $1/2\pi$ are even right! What then is $f(x)$, and why should we care about it? Physically, the result (10.1) states that the flow near the line of sources isn't quite purely two-dimensional, but also involves some axial flow; that is, ϕ has a non-zero gradient in the x direction. This would be true for non-constant $m(x)$ even if $f(x) \equiv 0$, but certainly the so-far unspecified function $f(x)$ influences this axial flow profoundly. It is time to add some more detail to the result (10.1).

Here is a formula for $f(x)$:

$$f(x) = -\frac{1}{4\pi} \int_{-\infty}^{\infty} m'(\xi) \operatorname{sgn}(x - \xi) \log 2|x - \xi| \, d\xi. \tag{10.2}$$

The non-trivial integro-differential nature of the result (10.2) reflects an important physical property, in that the relationship between the two functions $m(x)$ and $f(x)$ is non-local. That is, although the two dimensional flow in the cross-flow plane represented by the first term of (10.1) depends only on the local source strength at that plane x=constant, the axial flow generated by the term $f(x)$ depends also on the value of the source strength at all other stations ξ, weighted by the kernel function $\operatorname{sgn}(x - \xi) \log 2|x - \xi|$. An important feature of (10.2) is the fact that this kernel is an odd function of $(x - \xi)$. This has consequences such as guaranteeing validity of d'Alembert's paradox of zero drag, essentially because $\int m'(x) f(x) \, dx = 0$ for all $m(x)$. In some generalisations, e.g. to supersonic flow (Lighthill 1948), the kernel is not odd, and the drag is not zero.

How is (10.1) proved? There are probably many ways, but I know of two good ways. The easiest way (Thwaites 1960, p. 385) is to Fourier transform, noting that the Fourier transform of the unit source is proportional to the modified Hankel function $K_0(|k|r)$ where k is the Fourier variable. Then 'all' that is left to do is to expand this Bessel function for small r, using

$$K_0(|k|r) = -\log(|k|r) - \gamma + O(k^2 r^2 \log r) \tag{10.3}$$

(γ being Euler's constant 0.5772...), and fill in the details by inverting the Fourier transform.

There are lots of holes in the above summary, notably the fact that the Fourier variable k must be allowed to range to infinity, so the limit $|k|r \to 0$ is not as simple as just $r \to 0$. In my Ph.D. thesis (Tuck 1963), after much justified prodding by Ursell, I managed to plug some of these holes. In effect, rigorous completion of this derivation provides some smoothness restriction on $m(x)$. I was able to prove that (10.1) is valid uniformly with respect to x if $m(x)$ is continuous, at least once piecewise differentiable, and absolutely integrable over the whole x axis. However, this rigorous result did not have the full error estimate in (10.1) that seems to follow naturally from (10.3), but rather had the very conservative estimate $O(r^{1/5})$. No doubt a tighter error estimate could be proved.

In fact, I did not use the Fourier-transform method to prove this result. Instead I used a 'direct' method, one learned from Ursell, and which is I believe typical of Ursell's style. In order to motivate this, let us first ask why (10.1) is not an entirely trivial result analytically. After all, why not just naively expand the inverse square root as a Taylor series for small r? The answer is I hope obvious — the immediate result of letting r vanish on the left of (10.1) is a divergent integral, with a non-integrable singularity at $\xi = x$! Of course this had to be so, since after all the quantity on the right also becomes infinite when $r = 0$.

We circumvent this difficulty by excluding a 2δ neighbourhood of the singularity at $\xi = x$, writing

$$\int_{-\infty}^{\infty} = \int_{-\infty}^{x-\delta} + \int_{x-\delta}^{x+\delta} + \int_{x+\delta}^{\infty} \tag{10.4}$$

for some small number δ. Then in the remaining portions, we can indeed apply a Taylor series to the inverse square root. The result after an integration by parts is a formula which involves $\log \delta$. Not surprisingly, this blows up if we let $\delta \to 0$, since when there is no exclusion, the answer must diverge. We still have to estimate the excluded part, and this can also be done analytically, notably making use of the fact that when δ is small, then $m(\xi) \approx m(x)$ and this factor is approximately constant in the interval $(x - \delta, x + \delta)$. The remaining factor can be integrated exactly, and, lo and behold, when we let $\delta \to 0$, we get a term in $-\log \delta$ which cancels the corresponding term from the other two portions of the original integral, leaving behind the terms we really want involving $\log r$, etc.

When I first saw this work out, it seemed like almost unbelievable magic, and I urge others who delight in Ursell-style analysis to work through the details that I have only very roughly sketched here. There are other examples of asymptotic estimation of integrals where similar manipulations have value. It is remarkable how little needs to be said about δ; just that it is small compared to the scale for x-variations, but large compared to r.

In most applications, we are interested in cases where the range of integration is finite, i.e. where the source strength $m(x)$ vanishes identically outside some range $-\ell < x < \ell$. In that case, the result (10.1) can be derived rather more easily (Goldstein 1960, p. 184) simply by adding and subtracting $m(x)$ from the numerator. This leads to (10.1), but with

$$\begin{aligned} f(x) &= -\frac{1}{2\pi} m(x) \log \left[2\sqrt{\ell^2 - x^2} \right] \\ &+ \frac{1}{4\pi} \int_{-\ell}^{\ell} \frac{m(x) - m(\xi)}{|x - \xi|} \, d\xi, \end{aligned} \tag{10.5}$$

which reduces to (10.2) after a (very careful!) integration by parts, providing $m(\pm\ell) = 0$. Since we shall meet the integral in (10.5) again later, let us define a formal integral transform operator $\mathcal{S}$ by

$$\mathcal{S}m(x) = \int_{-\ell}^{\ell} \frac{m(x) - m(\xi)}{|x - \xi|} \, d\xi, \tag{10.6}$$

so that

$$f(x) = -\frac{1}{2\pi} m(x) \log \left[2\sqrt{\ell^2 - x^2} \right] + \frac{1}{4\pi} \mathcal{S}m(x). \tag{10.7}$$

Note that for differentiable $m(x)$, the integrand in the $\mathcal{S}$-transform is bounded, but has a step discontinuity at $\xi = x$. The operator $\mathcal{S}$ has several interesting properties (Tuck 1964a); for example (with $\ell = 1$)

$$\mathcal{S}P_n(x) = 2\sigma_n P_n(x), \tag{10.8}$$

where $P_n(x)$ is the Legendre polynomial and

$$\sigma_n = 1 + \frac{1}{2} + \frac{1}{3} + \ldots + \frac{1}{n} \tag{10.9}$$

($\sigma_0 = 0$). The $\mathcal{S}$-transform has not been studied very much, and there would seem to be scope for further investigation of the formal analytic properties of this operator in its own right.

The requirement that $m(\pm\ell) = 0$ for equivalence of (10.2) and (10.5) is a signal for consideration of a major sub-topic in slender body theory, namely that of end effects. Notice that unless this condition is met, then when considered on the whole x axis and vanishing identically outside $|x| < \ell$, the function $m(x)$ possesses (at least) step function discontinuities at $x = \pm\ell$, and hence violates the smoothness conditions apparently demanded for validity of (10.1). Using a rather artificial and probably unnecessary alternative formulation of slender body theory in terms of prolate spheroidal coordinates, I was in effect able to prove (Tuck 1964a) that even when such step discontinuities are present, (10.1) is still valid, providing (10.5) is used instead of (10.2) for $f(x)$.

10.2 Special cases

One of the first things one might like to use (10.1) for, is to prove an 'obvious' result, namely that a line source is a line of sources! Every beginning hydrodynamics student learns separately about point and line sources, but few will ever have seen the relationship between them proved. This is because it is not at all easy to prove it!

Suppose we take a constant source strength $m(x)$ =constant. Then the integral on the left of (10.1) diverges if the range of integration is doubly infinite, so we'd better assume a finite range $(-\ell, \ell)$, and ask what happens as $\ell \to \infty$. But that is just the same as asking what happens as $r \to 0$ at fixed ℓ, since ℓ provides the only comparison length scale. Now if we use (10.5) for $f(x)$ with $m(x)$ constant, the integral term in (10.5) is zero, and (10.1) becomes

$$-\frac{1}{4\pi} \int_{-\ell}^{\ell} \frac{m\, d\xi}{\sqrt{(x-\xi)^2 + r^2}} \approx \frac{m}{2\pi} \log r - \frac{m}{2\pi} \log \sqrt{1 - x^2/\ell^2}$$
$$- \frac{m}{2\pi} \log(2\ell). \qquad (10.10)$$

The last term still causes the right hand side to beome infinite in the limit as $\ell \to \infty$ (as it must, since the original integral on the left is divergent) but this is a strictly constant term, and can be ignored (or subtracted from the original integral) since constants are of no significance in velocity potentials. Then when we let $\ell \to \infty$, the second term on the right of (10.10) tends to zero so long as x is not close to $\pm\ell$. The resulting infinitely-long uniform-strength distribution of three-dimensional point sources has velocity potential $\frac{m}{2\pi} \log r$, which is the required two-dimensional line source potential.

An important hydrodynamic use of (10.1) is to solve the exterior Neumann problem of flow parallel to the axis of a fixed slender body of revolution, or equivalently the flow due to translational ('surge') motion of the body along its own axis. In that case, by using (10.1) in conjunction with the impermeability boundary condition on the surface of that body, it is possible to show (Ward 1955) that

$$m(x) = UA'(x), \qquad (10.11)$$

where $A(x)$ is the body's area of cross-section at station x and U is the speed of the x-directed uniform stream at infinity. Equation (10.11) is intuitively reasonable, since the body is pushing fluid aside in proportion to the rate at which its cross-section area is increasing.

A special case is that of a prolate spheroid of maximum radius $\epsilon\ell$, whose surface is

$$r = \epsilon\sqrt{\ell^2 - x^2}, \qquad (10.12)$$

and whose section area is

$$A(x) = \pi\epsilon^2\left(\ell^2 - x^2\right). \qquad (10.13)$$

Hence the slender body approximation suggests that this body will be generated by a line of sources of strength

$$m(x) = -2\pi U \epsilon^2 x \qquad (10.14)$$

proportional to x on the finite interval $(-\ell, \ell)$.

The distribution (10.14) is a typical example where there are step discontinuities in the source strength at the ends, and where we can expect end-effect singularities in the slender-body approximation. Nevertheless, (10.1) is valid subject to (10.5), and predicts that the disturbance velocity potential ϕ due to the spheroid (namely the quantity on the left of (10.1), with $m(x)$ given by (10.14)) is approximated for small ϵ by

$$\phi = -U\epsilon^2 \left[\log \left(\frac{r}{2\sqrt{\ell^2 - x^2}} \right) + 1 \right] x. \qquad (10.15)$$

Of course the flow past a spheroid can be solved exactly, for example (Lamb 1932, p. 139) in terms of a prolate spheroidal coordinate ζ (whose value is the sum of the distances from the foci divided by the distance between the foci), the solution being

$$\phi = U \lambda x, \qquad (10.16)$$

where

$$\lambda = -\frac{Q_1(\zeta)}{\zeta Q_1'(\zeta_0)}, \qquad (10.17)$$

with

$$Q_1(\zeta) = \frac{1}{2} \zeta \log \frac{\zeta + 1}{\zeta - 1} - 1 \qquad (10.18)$$

and where ζ_0 is the value of ζ on the body, namely

$$\zeta_0 = \left(1 - \epsilon^2 \right)^{-1/2}. \qquad (10.19)$$

A feature of this exact solution is that on the spheroid $\zeta = \zeta_0$ itself, the velocity potential is exactly a constant multiple $\lambda = \lambda_0$ of the uniform stream Ux. This property it clearly shares with the slender body approximation (10.15), and accuracy of the approximation

$$\lambda_0 = -\frac{Q_1(\zeta_0)}{\zeta_0 Q_1'(\zeta_0)} \approx -\epsilon^2 \left[\log \frac{1}{2} \epsilon + 1 \right] \qquad (10.20)$$

is an important measure of the accuracy of slender body theory. This comparison is not just of theoretical interest, since the ('surge') added mass of the spheroid can be shown to be equal to λ_0 times the mass of displaced fluid (Lamb 1932, p. 154, where $\lambda_0 = k_1$). Figure 10.1 compares the exact and slender body surge added mass coefficients. Slender body theory has better than 10% accuracy for about $\epsilon < 0.18$ and better than 1% for about $\epsilon < 0.05$. This suggests that it is of reasonable but not high accuracy for actual ship or projectile-like bodies. On the other hand, it

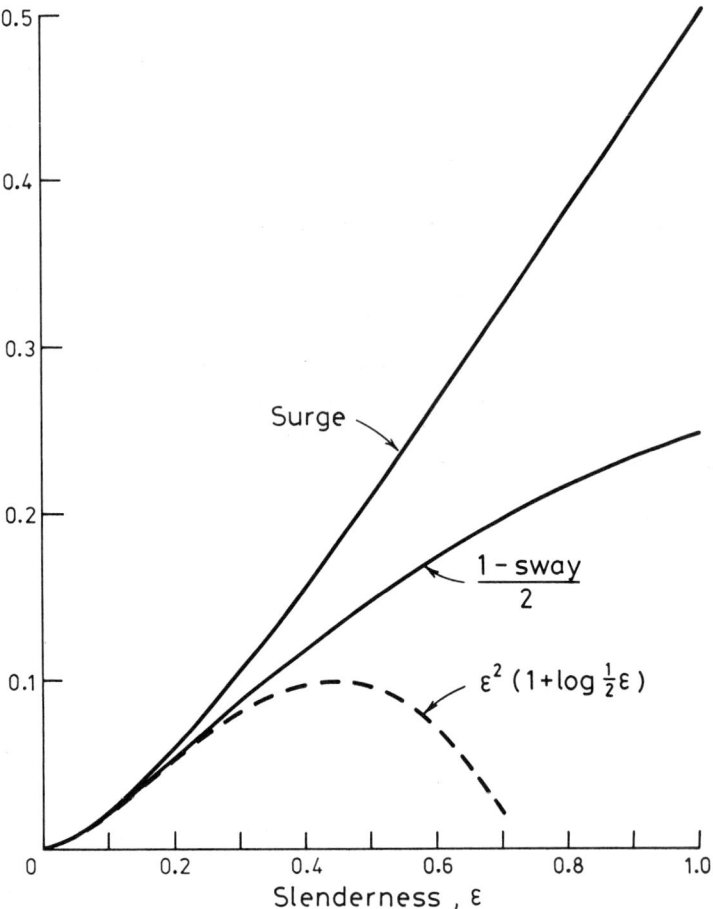

Figure 10.1: Surge and sway added masses of spheroids

ceases to give even qualitatively useful results for about $\epsilon > 0.4$, and of course fails totally to predict approach to the sphere added mass $\lambda_0 = 0.5$ at $\epsilon = 1$.

Another remarkable feature of the flow over spheroids is that in one sense the slender body result (10.14) is exact (see Chwang & Wu 1974). That is, if we ask what closed stream surface is generated by placing the linear point source distribution (10.14) in a uniform stream, the answer is (exactly) a spheroid. However, for not-so-slender spheroids, it is not exactly the spheroid (10.12). It extends slightly beyond the interval $(-\ell, \ell)$, and its maximum thickness is slightly different from $2\epsilon\ell$. Proving this result is one of the more difficult exercises that I often set for senior undergraduates, worthy of Ursell. Interestingly, the two-dimensional equivalent of this result (suggesting generation of flow over an elliptic cylinder by linear distributions of line sources) is *false*.

In the above connection, one may consider bodies generated by interior source distributions as generalised Rankine bodies, by analogy with the well known Rankine ovoid, obtained by placing a single discrete source and an equal sink in a stream. But in fact this is more than just an analogy. Suppose that (ignoring end effects for the moment) one asks the question: what is the prediction of slender body theory for flow parallel to the axis of a finite-length cylinder? Since the section area $A(x)$ of such a sharply truncated cylinder is a step function, slender body theory demands a source distribution that is a pair of Dirac delta functions, namely a discrete sink at the leading end and a discrete source at the trailing end. But this is just what generates the Rankine ovoid! In other words, a slender Rankine ovoid is a truncated cylinder. This conclusion is easy to accept if for example one computes and compares the exact shapes of ovoids and spheroids of equal small slendernesses ϵ (another of my favourite undergraduate exercises), as shown in figure 10.2 for the case $\epsilon = 0.2$. The increased severity of end effects is also shown by the fact

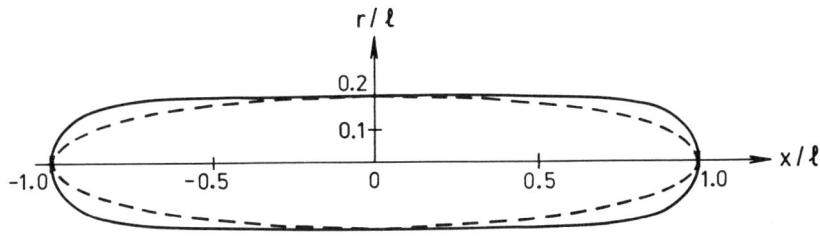

Figure 10.2: Spheroid (dashed) and Rankine ovoid with $\epsilon = 0.2$

that the ovoid extends a distance of the order of ϵl (comparable to its thickness) beyond the source sink pair, whereas the spheroid extends an asymptotically lesser distance of the order of $\epsilon^2 l$ beyond its linear source distribution.

10.3   Surface pressures

In axial flow over the body of revolution $r = R(x)$, slender body theory enables estimation of the pressure coefficient

$$C_p = \frac{p - p_\infty}{\frac{1}{2}\rho U^2} = 1 - \frac{q^2}{U^2}.$$ (10.21)

In the first place, the velocity magnitude can be approximated to give

$$C_p = -\frac{2}{U}\phi_x - \frac{1}{U^2}\phi_r^2,$$ (10.22)

where ϕ is the quantity on the left of (10.1). Then, on the body itself with $r = R(x)$, and using (10.1) subject to (10.7), we find

$$
\begin{aligned}
C_p(x) &= -(RR')'\log\frac{R^2}{4(l^2 - x^2)} - RR'\frac{2x}{l^2 - x^2} \\
&\quad - (R')^2 - \frac{d}{dx}\mathcal{S}RR',
\end{aligned}
$$ (10.23)

where $\mathcal{S}$ is the operator defined by (10.6). See Lighthill (1948) for a similar formula in supersonic flow.

The expression given in (10.23) is of order $\epsilon^2 \log \epsilon$, and has error of order $\epsilon^4 \log^2 \epsilon$. The fact that C_p is formally a small quantity is at first sight disturbing, since C_p should take the non-small value 1 at stagnation points. For spheroid-like bodies with ends of finite curvature where RR' is bounded at $x = \pm l$, the formula (10.23) predicts instead that C_p becomes infinite like the reciprocal of distance from the end. For sharper cone-like ends, such that R tends to zero linearly, there is a logarithmic infinity in the predicted pressure at the ends. For even sharper (cusped) ends there is no singularity — hardly surprising since there is no stagnation point either, and (10.23) correctly predicts then a uniform $O(\epsilon^2 \log \epsilon)$ pressure.

We can throw more light on this matter by considering the special case of the spheroid in detail. Then with $R(x)$ given by (10.12), we find from (10.23) that

$$C_p(x) = 2\epsilon^2\left(\log\frac{1}{2}\epsilon + 1\right) + \frac{\epsilon^2 x^2}{l^2 - x^2}$$ (10.24)

which is shown dashed for $\epsilon = 0.2$ in figure 10.3. The exact result can be written

$$C_p(x) = \frac{-2\lambda_0 - \lambda_0^2 + \epsilon^2 x^2/(l^2 - x^2)}{1 + \epsilon^2 x^2/(l^2 - x^2)}$$ (10.25)

where λ_0 is given by (10.20). Clearly (10.25) reduces to (10.24) (with the slender body approximation to λ_0) on neglect of $\lambda_0^2 = O(\epsilon^4 \log^2 \epsilon)$, and replacement of the denominator by 1. The solid curve in figure 10.3 gives the exact pressure at $\epsilon = 0.2$. Even for this not-particularly-slender body, the exact and approximate results are quite close for most x. The error at the mid-section $x = 0$ is about 14%, typical of slender body errors at this slenderness (cf. figure 10.1). What appears in the analysis to be a dramatic end-effect error, with a conflicting prediction of infinity by (10.24), and 1 by (10.25), is shown in a different light by the plots. The point is that when slender body theory predicts an $O(\epsilon^2 \log \epsilon)$ pressure coefficient, a value of infinity for that prediction is quite a good estimate of the (formally large on $\epsilon^2 \log \epsilon$ scale) exact stagnation pressure $C_p = 1$.

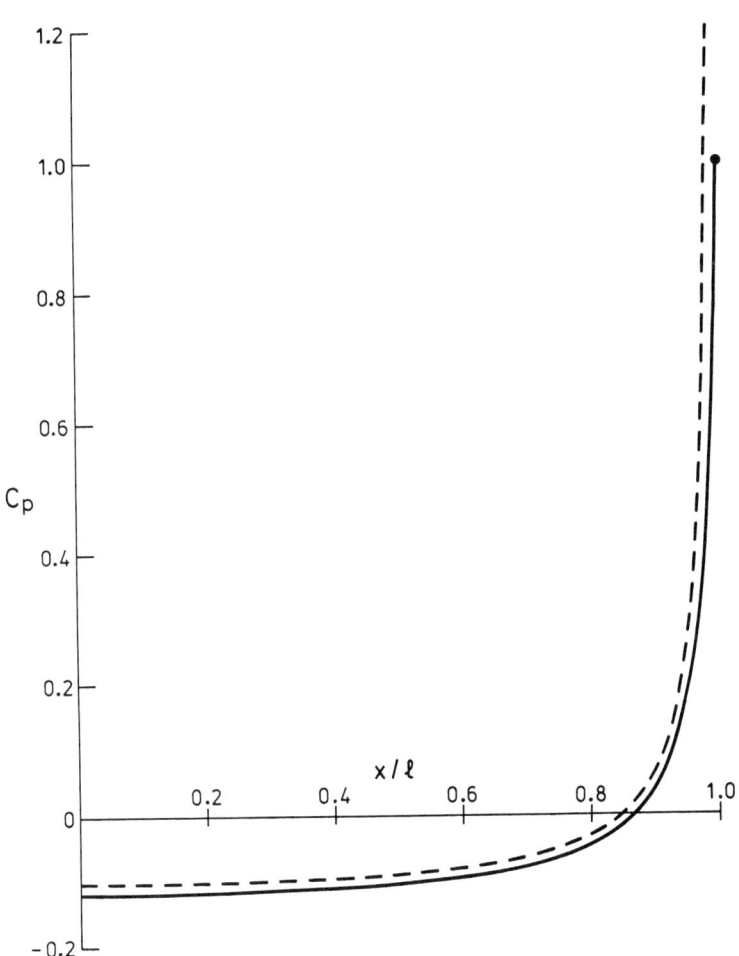

Figure 10.3: Exact and slender body (dashed) pressure for a spheroid with $\epsilon = 0.2$

We can however do better if we really need good end accuracy, using a composite approximation (Van Dyke 1975, p. 94) which retains validity near the ends by matching with a local flow there. In the special case of the spheroid, this simply corresponds to inconsistently *not* replacing the denominator of (10.25) by 1, while still neglecting λ_0^2. In the general case, the composite formula will involve a multiplicative factor containing in its denominator an expression like that in (10.25). The result is similar to what one obtains by Lighthill's (1949) technique. The composite result for general bodies is also derived in Tuck (1964a; equation 3.15 where a misprint of omission of a term '$+\eta_0^2$' from inside the square bracket needs to be corrected) using a formulation in terms of spheroidal coordinates which automatically preserves uniform validity near the ends because the approximation is not one of small r but rather of small values of a spheroidal coordinate. If this composite procedure is tested on spheroids at $\epsilon = 0.2$, all that happens is that there is a rapid crossover between the dashed and solid curves of figure 10.3 at about $x/\ell = 0.9$; for smaller ϵ, this crossover occurs nearer the end. Except for some special applications such as that demanding boundary layer computations with high accuracy near the ends, it seldom seems worthwhile making end-effect corrections.

In the general case, the slender body formula (10.24) can be used for various $R(x)$, and tends to produce small negative pressures over most of the body, with rapid rises to the positive-infinite end singularity as $x \rightarrow \pm\ell$. The end-effect issues discussed above for bodies with ends like spheroids become more difficult to resolve for blunter bodies. At an extreme, one might like to apply the formula to truncated cylinders $R =$ constant, i.e. to the slender body limit of Rankine ovoids. Although validity of (10.24) for such bodies is questionable, it does suggest a character to the pressure distribution that is present in the exact results, namely a singularity of δ' character at the ends. This is shown by the exact C_p for a Rankine ovoid of slenderness 0.2, shown in figure 10.4 and compared to the (dashed) spheroid pressure at the same slenderness. The sharp pressure dips are typical of blunt-ended bodies, and are of considerable importance in applications. It is an interesting analytic exercise to verify d'Alembert's paradox using the formula (10.24). That is (subject to suitable end-smoothness restrictions if the range of integration is finite) it is possible to prove that

$$\int R(x)R'(x)C_p(x)\,dx = 0 \tag{10.26}$$

for arbitrary $R(x)$. Another use of (10.24) is to compute the non-zero 'lift' force F on half of the body (sliced by a plane containing its axis, which can be taken as a ground plane). This has applications to automobile or train aerodynamics, and if one makes plausible assumptions about the pressure in the small gap between vehicle and ground (Tuck 1975, p. 133), one finds the simple formula

$$F = -\rho U^2 \int R'(x)^2 R(x)\, dx,\qquad (10.27)$$

the negative sign implying a force directed toward the ground.

10.4 Higher approximations

It is in principle possible to improve the slender body approximation, viewing it as the leading term in a formal asymptotic series for small slenderness. One obvious way to derive such a series is simply to use more terms in the Bessel function expansion begun by (10.3), and the resulting series generalising (10.1) is

$$
\begin{aligned}
&-\frac{1}{4\pi}\int_{-\infty}^{\infty}\frac{m(\xi)\, d\xi}{\sqrt{(x-\xi)^2 + r^2}}\\
&= \sum_{n=0}^{\infty}\frac{(-r^2/4)^n}{(n!)^2}\frac{\partial^{2n}}{\partial x^{2n}}\left[\frac{1}{2\pi}m(x)(\log r - \sigma_n) + f(x)\right],\quad (10.28)
\end{aligned}
$$

where σ_n is given by (10.9).

I mention in passing another way, one which Ursell has often used (e.g. Ursell 1962), namely to take an immediate Laplace transform with respect

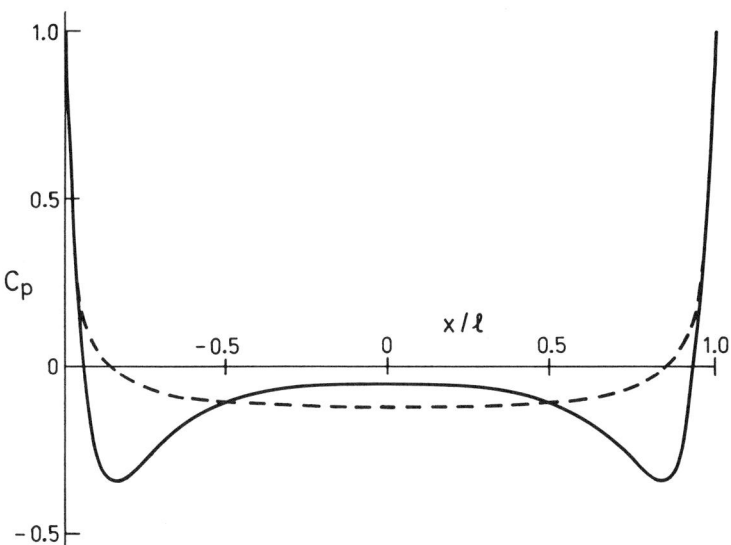

Figure 10.4: Surface pressures on the bodies of figure 10.2

to r as a 'time-like' variable. Then if p is the Laplace-transform variable, it is often possible to perform a formal large-p asymptotic expansion of the resulting Laplace-transformed integral, then invert the Laplace transform term by term. In doing this, a useful guide is that each term must itself be harmonic.

Commonly we may be very content with just the first or second correction to (10.1). Retaining only two terms in the series (10.28) gives

$$-\frac{1}{4\pi}\int_{-\infty}^{\infty}\frac{m(\xi)\,d\xi}{\sqrt{(x-\xi)^2+r^2}} = \frac{1}{2\pi}m(x)\log r + f(x)$$

$$-\frac{1}{4}r^2\left[\frac{1}{2\pi}m''(x)(\log r - 1) + f''(x)\right] + O(m''''r^4\log r). \quad (10.29)$$

Notice that although the error term is formally of the order of $r^4\log r$, it also involves fourth derivatives of $m(x)$ and there are error terms of the order r^4 involving fourth derivatives of $f(x)$. Hence there are more stringent demands placed upon the smoothness of the function $m(x)$ in order that the error be uniformly of this order of smallness, demands that are unlikely to be met for finite-length slender bodies.

Indeed, extending (10.1) to more terms as a series in r for a given source strength function $m(x)$ is only a small part of the task of constructing a higher-order slender body theory. If we are solving an axisymmetric exterior Neumann problem, for example, it is also necessary to improve the degree of accuracy with respect to which the body boundary condition is satisfied, so that the estimate (10.11) of $m(x)$ in terms of body geometry must be improved simultaneously.

For a body of revolution $r = R(x)$ with section area $A(x) = \pi R(x)^2$, a second-order improvement of (10.11) is (Tuck 1963)

$$m(x) = U\frac{d}{dx}\left\{A(x)B(x)\right\}, \quad (10.30)$$

where

$$B(x) = 1 + \frac{1}{2\pi}A''(x)(\log R(x) - \frac{1}{2}) + \frac{f'(x)}{U}.$$

If ϵ measures slenderness, so that $A = O(\epsilon^2)$, the leading-order source strength as given by (10.11) is $O(\epsilon^2)$, the second-order correction contained in (10.30) is $O(\epsilon^4\log\epsilon)$, and the error in (10.30) is $O(\epsilon^6\log^2\epsilon)$. Note the presence of the 'interaction' contribution in $f'(x)$, which means that at second order the source strength is non-local, the value at station x depending on the geometry at all other stations ξ.

Once again, the presence of two extra derivatives with respect to x in the correction term signals that smoothness demands will be severe, and hence, for finite length bodies, that end-effect difficulties will arise. If the ends happen to be spheroid-like in axisymmetric flow, these difficulties can be eliminated by simultaneously improving the estimate of the 'stand-off' distance, namely the (small) extent to which the body extends beyond the line segment on which we distribute sources and sinks. Nothing seems to have been done about end-effect corrections for ends that are either more or less blunt than spheroids, or for bodies not of revolution or in non-axisymmetric flow.

10.5   Lateral flow

A particular non-axisymmetric flow of interest is that perpendicular to the axis of a slender body, say in the y-direction. An analytic result that is needed is that for the y derivative of (10.1), where $r^2 = y^2 + z^2$, namely

$$\frac{\partial}{\partial y}\left[-\frac{1}{4\pi}\int_{-\infty}^{\infty}\frac{m(\xi)\,d\xi}{\sqrt{(x-\xi)^2+r^2}}\right] = \frac{1}{2\pi}m(x)\frac{y}{r^2}+O(r\log r).\quad(10.31)$$

This result simply says that a line distribution of three-dimensional lateral point dipoles behaves locally as a two-dimensional line dipole, which is a natural generalisation of the corresponding source result.

But it says more than that, in view of the fact that the term $f(x)$ has disappeared from (10.1) upon y-differentiation. There are no 'interaction between section' effects in (10.31), in contrast to (10.1). The velocity potential at station x depends only on the dipole strength at that station, and hence ultimately only on the local geometry of the slender body at that station.

This fundamental difference between lateral dipole and source distributions has profound consequences when the full slender body problem is solved by approximation to the body boundary condition. Whether or not the slender body is a body of revolution, what it means is that a purely two-dimensional or 'strip' theory, in which all axial flow is neglected, and the flow is replaced at every section by a two-dimensional flow over a cylinder with the local cross-section, is the correct first slender body approximation in lateral flow.

This is a result whose consequences extend deep into many generalised applications of slender body ideas. It is both a blessing and a curse. The simplicity of the stripwise two-dimensional slender body theory is sometimes useful. But at the same time we lose some of the more interesting physical features of interaction between sections via axial flow. For example, there can be no effects of transverse waves on lateral forces on slender bodies (to leading order in slenderness).

Another example is that the ('sway') added mass of a body of revolution in lateral flow is to leading order in slenderness just equal to the displaced mass of fluid, as would be the case for a circular cylinder. This is inevitable since it is true section by section, and we only have to add up the individual contributions. If we want an improved estimate, we have to, among other things, y-differentiate (10.30) instead of (10.1), yielding

$$\frac{\partial}{\partial y}\left[-\frac{1}{4\pi}\int_{-\infty}^{\infty}\frac{m(\xi)\,d\xi}{\sqrt{(x-\xi)^2+r^2}}\right] = \frac{1}{2\pi}m(x)\frac{y}{r^2}$$
$$-\frac{1}{4\pi}m''(x)y(\log r - \frac{1}{2})$$
$$-\frac{1}{2}yf''(x) + O(r^3\log r)$$

The interesting term is the last one, showing that interaction between sections does indeed occur in lateral slender body theory, but only at second order in slenderness.

After considerable effort, it is possible to use this result to derive the following estimate for the (sway) added mass M of a rigid body of revolution of a general shape $r = R(x)$, namely

$$\frac{M}{\rho V} = 1 + \frac{1}{\pi V}\int_{-\ell}^{\ell}A'(x)C(x)\,dx, \tag{10.32}$$

where

$$C(x) = A'(x)\log\left(\frac{R(x)}{2\sqrt{\ell^2-x^2}}\right) + \frac{1}{2}\mathcal{S}A'(x)$$

and $V = \int_{-\ell}^{\ell}A(x)dx$ is the displaced volume. That is, the sway added mass is equal to the $O(\epsilon^2)$ displaced mass ρV, plus a correction of the order $\epsilon^4\log\epsilon$, which involves three-dimensional effects and interactions between sections via the same integral transform operator $\mathcal{S}$ that we met in (10.6).

In particular, for the spheroid (10.12),

$$\frac{M}{\rho V} = 1 + 2\epsilon^2\left[\log\frac{1}{2}\epsilon + 1\right]. \tag{10.33}$$

This can be compared (see Figure 1) to the exact result (Lamb 1932, p. 154, where this ratio is called k_2)

$$\frac{M}{\rho V} = \frac{1-(\zeta_0^2-1)Q_1(\zeta_0)}{1+(\zeta_0^2-1)Q_1(\zeta_0)}, \tag{10.34}$$

where ζ_0 and Q_1 are as in (10.19), (10.18). The two-term small-ϵ expansion of (10.34) agrees with (10.33). Note that, comparing (10.33) with (10.20), to leading order in slenderness, the three-dimensional correction to the sway added mass is simply to *subtract twice the surge added mass*. Figure 1 confirms this, and suggests that slender body theory is somewhat more accurate for lateral flow than for axial flow.

The above can easily be generalised to non-rigid bodies, allowing estimation of important three-dimensional effects on bending vibrations ('springing') of elongated bodies like ships. Lighthill (1948) obtained a result like (10.32) for the lift in supersonic flow.

10.6   More generalisations

There are many applications and generalisations of slender body theory, in which the ideas mentioned already play a role, but not necessarily a direct one. Compressibility effects in steady aerodynamics (both subsonic and supersonic, Ward 1955) and wave effects in steady ship hydrodynamics (Tuck 1964b) are examples where the generalisation is direct. That is, the problem is still solved by line distributions of sources, and a result like (10.1) still applies, but the integral relationship between $f(x)$ and the source strength $m(x)$ differs from (10.2). An interesting subtlety about the ship hydrodynamic problem is that the error in (10.1) is no longer as small as $O(r^2 \log r)$, but rather is formally $O(r \log r)$. A second-order theory therefore applies with some $O(r \log r)$ terms due to wave effects appearing before the need to worry about terms like those in (10.29).

Some other generalisations do not directly involve the asymptotic result (10.1), but do involve the integral relationship (10.5), and hence the integral operator S defined by (10.6). For example, slow viscous (Stokes) flow about the fixed no-slip slender body of revolution $r = R(x), |x| < \ell$ demands solution of the integral equation

$$m(x) + 2\pi U + 2m(x) \log \left(\frac{R(x)}{2\sqrt{\ell^2 - x^2}} \right) + Sm(x) = 0 \qquad (10.35)$$

for a 'Stokeslet' strength $m(x)$. This was derived almost parenthetically in Tuck (1964a) and expanded upon slightly in Tuck (1968), but the subject of slender body theory at low Reynolds number then took off in many directions, mostly stimulated by biomechanical applications, see e.g. Blake & Sleigh (1974).

Similarly, the deformation of an elastic half-space $z > 0$ by a slender punch which displaces the contact zone $|y| < R(x), |x| < \ell$ to $z = w(x)$ requires solution of the integral equation

$$\frac{2\pi\mu}{1-\nu} w(x) + 2m(x) \log \left(\frac{R(x)}{4\sqrt{\ell^2 - x^2}} \right) + Sm(x) = 0 \qquad (10.36)$$

for the net vertical force $m(x)$ at each station x, where μ, ν are elastic constants. This result was derived by Kalker (1972), and others, see Mei & Tuck (1983).

The integral equations (10.35) and (10.36) are remarkably similar to each other, and to expressions like (10.32). The recurring theme of occurrence of this type of integral operator S in slender body theory does not seem to have been remarked upon before. As integral equations with $m(x)$ as unknown, (10.35) and (10.36) appear to be rather unpleasant in spite of their linearity. In both cases there are violent oscillatory end effects unless the body is a spheroid (when there is an exact solution), and the elimination of these end effects seems to be a difficult numerical problem. However, an attractive alternative interpretation for both

(10.35) and (10.36) is in terms of an inverse problem with $m(x)$ given, in which case the body's shape $R(x)$ can be determined explicitly.

Slender bodies need not be straight. That is, a snake-like body with slenderness ϵ is in general such that when $\epsilon \to 0$, it collapses to a curve in three-dimensional space. If that curve happens to be a straight line segment, the results reduce to what we have been discussing so far. If not, the flow is modelled in the far field by a distribution of sources and dipoles along the limiting curve. If we then move into the near field in order to apply boundary conditions, we need the formal limit as a local polar coordinate approaches zero, in a plane which is locally normal to the limiting curve. There are many analytical difficulties with this limit, see Tuck (1975, p. 130).

Slender bodies also need not be bodies of revolution. However, it seems to me that the only sensible way to discuss slender bodies with a general cross-section is via the method of matched asymptotic expansions (as used for the ship problem in Tuck 1964b; see also Tuck 1989) in which there is a non-trivial 'inner' boundary-value problem for the two-dimensional Laplace equation, to be solved at each section, then matched with a singularity distribution in the 'outer' region. The analysis needed to match this outer problem is similar to that reviewed here, since what is needed is the limit as the axis of the singularities is approached. However, in view of my intention to avoid intuitive concepts like matched expansions in this article, this does appear to be a suitable place to stop.

10.7   References

Blake, J.R. & Sleigh, M.A. 1974 Mechanics of ciliary locomotion. *Biol. Rev.* **49**, 85–125.

Chwang, A.T. & Wu, T.Y. 1974 A note on the potential flow involving prolate spheroids. *Schiffstechnik* **21**, 19–31.

Goldstein, S. 1960 *Lectures on fluid mechanics.* New York: Interscience.

Kalker, J.J. 1972 On elastic line contact. *J. appl. Mech.* **39**, 1125–1132.

Lamb, H. 1932 *Hydrodynamics.* New York: Dover.

Lighthill, M.J. 1948 Supersonic flow past slender pointed bodies of revolution at yaw. *Q. Jl Mech. appl. Math.* **1**, 77–89.

Lighthill, M.J. 1949 A technique for rendering approximate solutions of physical problems uniformly valid. *Phil. Mag.* Ser. 7 **40**, 1179–1201.

Mei, C.C. & Tuck, E.O. 1983 Contact of one or more slender bodies with an elastic half space. *Int. J. Solids Struct.* **19**, 1–23.

Thwaites, B. 1960 *Incompressible aerodynamics.* Oxford.

Tuck, E.O. 1963 *The steady motion of a slender ship.* Ph.D. thesis, University of Cambridge.

Tuck, E.O. 1964a Some methods for blunt slender bodies. *J. Fluid Mech.* **18**, 619–635.

Tuck, E.O. 1964b A systematic asymptotic expansion procedure for slender ships. *J. Ship Res.* **8**, 15–23.

Tuck, E.O. 1968 Toward the calculation and minimization of Stokes drag on bodies of arbitrary shape. *3rd Austral. Conf. Hydraul. Fluid Mech.*, Nov. 1968. Proc. Institution Engineers Australia, 1970, pp. 29–32.

Tuck, E.O. 1975 Matching problems involving flow through small holes. *Adv. Appl. Mech.* **15**, 89–158.

Tuck, E.O. 1989 A submerged body with zero wave resistance. *J. Ship Res.* **33**, 81–83.

Ursell, F. 1962 Slender oscillating ships at forward speed. *J. Fluid Mech.* **14**, 496–516.

Van Dyke, M. 1975 *Perturbation methods in fluid mechanics.* Parabolic.

Ward, G.N. 1955 *Linearised theory of high-speed flow.* Cambridge.

11

Vertical barriers, sloping beaches and submerged bodies

D.V. Evans
University of Bristol

*As a one-time student, postgraduate and colleague of Fritz
Ursell at Manchester, and a life-time admirer of his work,
it is a great pleasure for me to contribute a paper to this
Volume. What I shall endeavour to do is to discuss certain
problems I have worked upon over the years which have been
directly influenced by some previous paper or papers by
Fritz. I shall concentrate on three areas; the interaction of
waves with vertical barriers, sloping beaches, and submerged
bodies. In particular I shall consider extensions and
applications of his explicit vertical barrier solution of 1947,
his edge wave solution of 1952 and his submerged cylinder
papers of 1950 and 1951. It is a mark of the importance of
these papers that they continue to attract interest and
attention 40 years on. Finally I shall describe some recent
work on trapped modes in acoustic wave-guides or wave
channels containing an obstruction which originated in a
remark made by Fritz during a recent visit to Bristol.*

11.1   Vertical barriers

Explicit solutions to the linearised equations for water waves
when rigid boundaries are present are extremely rare. However Havelock
(1929), by extending Fourier's integral formula, showed how the velocity
potential could be determined everywhere when the horizontal velocity
was prescribed on a vertical plane wall extending throughout the fluid
and making small horizontal harmonic oscillations. The solution, for both
finite and infinite depth water, was also extended to finding the potential
due to a prescribed radial velocity of a vertical circular cylinder extending throughout the depth. These solutions have been used extensively
ever since but the first person to exploit the usefulness of the Havelock
wavemaker solution was Ursell (1947) in considering the two-dimensional
problem of the scattering of a plane wave incident upon a fixed rigid
barrier extending a depth a into infinitely deep water.

The problem reduces to finding the velocity potential

$$\Phi(x, y, t) = \mathrm{Re}\{\phi(x, y)e^{-i\omega t}\}$$

where ω is the radian frequency and $\phi(x, y)$ satisfies

$$\nabla^2 \phi(x, y) = 0 \quad \text{in the fluid} \tag{11.1}$$

$$K\phi + \phi_y = 0, \quad y = 0, \quad K = \omega^2/g \tag{11.2}$$

$$\phi_x = 0, \quad x = 0, \quad 0 < y < a \tag{11.3}$$

$$\phi, \nabla \phi \to 0, \quad y \to \infty \tag{11.4}$$

$$\phi \sim \begin{cases} \phi_0 \left\{ e^{-iKx - Ky} + Re^{iKx - Ky} \right\}, & x \to +\infty \\ \phi_0 T e^{-iKx - Ky}, & x \to -\infty. \end{cases} \tag{11.5}$$

Earlier, Dean (1945), using complex variable methods, had solved for a submerged vertical barrier extending upwards to a distance a beneath the surface, and it was already becoming clear that problems involving thin vertical rigid boundaries could be solved explicitly using the so-called reduction technique. Later, Lewin (1963), Mei (1966) and Porter (1974) provided general results for such cases.

The reason for the success of the method lies in the fact that if we introduce a complex potential $w(z)$ whose real part is $\phi(x, y)$ then, if

$$W(z) = \frac{dw}{dz} + iKw,$$

$$\mathrm{Im}\, W(z) = 0, \quad z \text{ real}, \quad x \neq 0$$

whilst the condition $\phi_x = 0$ on a vertical thin barrier, carries over to $\mathrm{Re}\, W(z) = \text{constant}$ on the barrier. Standard methods exist for finding $W(z)$ for different combinations of barriers and $w(z)$ is then found by integration and the application of the far-field conditions.

Armed with the fact that an explicit solution existed, Ursell chose to use Havelock's wavemaker solution to derive a (real) integral equation for the unknown horizontal velocity beneath the fixed barrier. Application of the operator $K + (d/dy)$ reduced this equation to a form occurring in aerofoil theory for which a simple solution existed. The difficult part was still to come and Fritz provided us with a first indication of his considerable analytical skills in recovering the potential $\phi(x, y)$.

The full solution is

$$\phi(x, y) = \phi_{\mathrm{inc}}(x, y) + \psi(x, y) \tag{11.6}$$

where

$$\phi_{\mathrm{inc}}(x, y) = \{\pi I_1(Ka) + iK_1(Ka)\} \exp\{-iKx - Ky\},$$

for all x, and

$$\psi(x,y) = \pi I_1(Ka)\exp\{iKx - Ky\}$$
$$+ \int_0^\infty \frac{k\cos ky - K\sin ky}{k^2 + K^2}J_1(ka)e^{-kx}\,dk$$

for $x > 0$, where

$$\psi(-x,y) = -\psi(x,y)$$

determines the solution in $x < 0$. Here I_1, K_1 are modified Bessel functions and the incident wave amplitude is clearly arbitrary.

The solution (11.6) should be treasured — it is a rare explicit solution and is only matched by a similar solution for the infinite submerged barrier of Dean (1945) obtainable by similar methods. But our enjoyment of (11.6) has a utilitarian as well as aesthetic side to it. We shall see that we can put (11.6) to work in a variety of ways.

Suppose we try and solve the next most difficult problem, the scattering of an *obliquely* incident plane wave by such a barrier, assuming the barrier to extend indefinitely in the horizontal z-direction. We write

$$\Phi = \mathrm{Re}\,\phi(x,y)\exp\{ilz - i\omega t\}$$

where $l = K\sin\alpha$ and seek a potential ϕ describing the scattering of an incident plane wave making an angle α with the x-y plane. Then all conditions on ϕ remain the same except that (11.1) is replaced by

$$(\nabla^2 - l^2)\phi(x,y) = 0. \tag{11.7}$$

Now complex variable theory is not obviously applicable and the Havelock wavemaker approach gives an integral equation which, after reduction, cannot be inverted explicitly.

What we can do, however, is to construct integral equations for the horizontal velocity under, and the jump in the potential across, the barrier, of the forms

$$\left.\begin{array}{rll} \displaystyle\int_a^\infty f(t)L(y,t)dt &=& e^{-Ky}, \qquad y > a, \\[2mm] \displaystyle\int_a^\infty f(t)e^{-Kt}dt &=& A \equiv -\dfrac{iT\cos\alpha}{\pi R}, \end{array}\right\} \tag{11.8}$$

and

$$\left.\begin{array}{rll} \displaystyle\int_0^a g(t)M(y,t)dt &=& e^{-Ky}, \qquad 0 < y < a, \\[2mm] \displaystyle\int_0^a g(t)e^{-Kt}dt &=& (\pi^2 K^2 A)^{-1}, \end{array}\right\} \tag{11.9}$$

where R, T are to be determined and A is real.

Both of the systems (11.8) and (11.9) are of the form

$$Lf = h, \qquad (f, h) = A \qquad (11.10)$$

where L is an integral operator with the property $(Lg, g) > 0$, $g \neq 0$ and

$$(Lf, g) = (f, Lg). \qquad (11.11)$$

Now it is straightforward to show (Jones 1964, pp. 269–271) that if $f_1 = a_1 \psi$ approximates f to $O(\epsilon)$ then

$$A_1 = \frac{(\psi, h)^2}{(L\psi, \psi)} \leq A \qquad (11.12)$$

approximates A to $O(\epsilon^2)$. The same is true for $g_1 = b_1 \chi$ but now since A appears in the denominator in (11.9)

$$A_2 = \frac{(\chi, M\chi)}{\pi^2 K^2(\chi, h)^2} \geq A. \qquad (11.13)$$

We therefore have complementary bounds for A whose values depend upon finding a good one-term approximation to $f(y)$ or $g(y)$. Ideally such an approximation would share as many properties of the true solution as possible. What better choice than the Ursell solution (11.6) which is, of course, exact, when $\alpha = 0$.

A certain amount of manipulation is necessary to evaluate (11.12), (11.13) using $\psi(y)$ and $\chi(y)$ derived from (11.6). Finally,

$$\frac{1}{A_1} = \frac{\pi^2}{K_1^2(\mu)} \int_0^\infty \frac{t J_1^2(t) dt}{(t^2 + \mu^2)(t^2 + \mu^2 \sin^2 t)^{\frac{1}{2}}} \qquad (11.14)$$

$$A_2 = \frac{1}{\pi^2 I_1^2(\mu)} \int_0^\infty \frac{(t^2 + \mu^2 \sin^2 \alpha)^{\frac{1}{2}} J_1^2(t)}{t^2 + \mu^2} dt, \qquad (11.15)$$

where $\mu = Ka$ and $A_1 \leq A \leq A_2$. Clearly therefore, since $1 - R = T, |R_1| \leq |R| \leq |R_2|$ from (11.8) where R_i $(i = 1, 2)$ are approximations to the true reflection coefficient R. For $\alpha = 0$, since

$$\int_0^\infty \frac{J_1^2(t)}{t^2 + \mu^2} dt = I_1(\mu) K_1(\mu),$$

$$A_1 = A_2 = A = \frac{K_1(\mu)}{\pi I_1(\mu)}$$

and

$$R = \frac{\pi I_1(\mu)}{\{\pi I_1(\mu) + i K_1(\mu)\}} \qquad (11.16)$$

in agreement with (11.6).

For general α, as table 11.1 illustrates, the approximation is good, even for $\alpha = 75°$. The maximum relative error in $|R|$ is about 2% using this approximation.

	$\alpha = 0$	$\alpha = 30°$		$\alpha = 60°$		$\alpha = 75°$	
Ka	$|R_i|$	$|R_1|$	$|R_2|$	$|R_1|$	$|R_2|$	$|R_1|$	$|R_2|$
0.2	0.0660	0.0569	0.0569	0.0326	0.0326	0.0168	0.0168
0.4	0.2816	0.2432	0.2430	0.1402	0.1396	0.0726	0.0722
0.6	0.6033	0.5389	0.5382	0.3353	0.3327	0.1791	0.1771
0.8	0.8447	0.7971	0.7961	0.5861	0.5806	0.3457	0.3401
1.0	0.9470	0.9252	0.9246	0.7965	0.7906	0.5557	0.5458
1.4	0.9934	0.9900	0.9898	0.9652	0.9629	0.8808	0.8721
1.8	0.9990	0.9984	0.9984	0.9942	0.9936	0.9775	0.9747

Table 11.1: Estimates of the reflection coefficient for the scattering of a plane wave at oblique incidence to a vertical barrier. Note that $|R_1| = |R_2|$ when $\alpha = 0$.

The method can be used to estimate $|R|$ for two vertical barriers, with $\alpha = 0$ and good accuracy is achieved provided the barriers are not too close together. In addition because of the complementary bounds the method can be used to prove that there exist discrete spacings and frequencies for which $|R| = 1$ and an incident wave is totally reflected. Details are given in Evans & Morris (1972 a,b).

Another illustration of the value of the Ursell solution is in predicting the waves radiated to either infinity by an arbitrary velocity of the barrier. There is now no incident wave but we have to satisfy the condition

$$\phi_x = f(y), \quad x = 0, \quad 0 < y < a.$$

We seek the complex number A where

$$\phi \sim A \operatorname{sgn} x \, e^{iK|x| - Ky}, \quad |x| \to \infty.$$

A simple application of Green's theorem to the Ursell solution and $\phi(x, y)$ results in

$$A = \frac{-2ia^{-1}}{\pi I_1(\mu) + i K_1(\mu)} \int_0^a \frac{y e^{Ky}}{\sqrt{a^2 - y^2}} \int_0^y f(t) e^{-Kt} \, dt \, dy.$$

For example when the barrier rolls about the point $x = 0$, $y = b$, $0 \le b \le a$ with amplitude θ_0,

$$A = \frac{-\omega \theta_0 \pi}{\mu \{\pi I_1(\mu) + i K_1(\mu)\}} \left\{ \frac{1}{2} - \frac{(1 - Kb)}{Ka}(I_1(\mu) + L_1(\mu)) \right\}$$

where L_1 is a Struve function, in agreement with Ursell (1948); see Evans (1976a).

As a final example of the usefulness of the explicit barrier solution we consider a problem of wave energy extraction.

The idea is to model an oscillating water column device in two dimensions by means of two closely-spaced parallel vertical plates immersed to a depth a, a distance $2b$ apart. The dynamic pressure due to the incident waves at the opening drives the slug of fluid between the plates backwards and forwards and drives a simple spring and dashpot system installed at the other end. The method of matched asymptotic expansions can be used on this problem. Far away from the plates the plates appear as one and the problem reduces to the scattering of an incident wave by a single plate in the presence of a pulsating source of unknown strength at its tip. Thus the solution is just the sum of the Ursell potential and a wave source. In the inner region the flow appears to be simply potential flow through two semi-infinite plates. Matching in an overlap region provides an estimate for the efficiency of wave-energy absorption on the assumption of plate spacing small compared to incident wavelength. Details are given in Evans (1978) where it is confirmed that the highest possible efficiency is 50%, a well-known result for such problems. The technique was first used in this context by Newman (1974) who used it to confirm the existence of zero transmission for particular configurations of two closely-spaced partly submerged barriers.

It should be clear from the above illustrations that the Ursell exact solution for the scattering of waves by a fixed partly immersed vertical barrier has proved extremely valuable in deriving approximate results in more difficult problems.

11.2   Sloping beaches

In 1846 Stokes produced a solution to the linearised water wave equations which described a wave-train which travelled unchanged over a uniform sloping beach, of angle β to the horizontal, in a direction parallel to the shoreline, but which decayed exponentially out to sea. The solution only existed for a particular relation between its radian frequency ω and its long-shore wavenumber k, namely

$$\omega^2/g = K = k \sin \beta. \tag{11.17}$$

Clearly this requires $k > K$ and it was not until over a century later that the solution for the scattering of an obliquely incident sinusoidal wavetrain by a sloping beach was obtained (see, for example, Peters (1952), Roseau (1958)). This solution is valid for all $k < K$ and we see therefore that for given k, the frequency of the Stokes edge-waves, as they are called, given by (11.17) is a discrete frequency lying below the continuous set of frequencies ω, which satisfy $\omega^2 > gk$ for which a solution to the scattering problem exists.

In a remarkable paper Ursell (1952) showed that the Stokes edge wave was just one of a set of $N + 1$ bounded trapped modes each with its own discrete frequency ω, where

$$\omega^2/g = k \sin (2n + 1)\beta, \quad n = 0, 1, \ldots, N \tag{11.18}$$

and N is the largest integer for which $(2N + 1)\beta < \pi/2$.

The conditions satisfied by the velocity potential

$$\Phi(x,y,z,t) = \operatorname{Re} \phi(x,y)e^{i(kz-wt)}$$

are

$$(\nabla^2 - k^2)\phi = 0 \quad \text{in the fluid} \tag{11.19}$$

$$\phi_n = 0 \quad \text{on the sloping beach} \tag{11.20}$$

$$K\phi - \phi_y = 0, \quad y = 0, \quad x > 0 \tag{11.21}$$

$$\phi \to 0 \text{ as } r \to \infty \text{ in the fluid.} \tag{11.22}$$

Here x, y, z are Cartesian co-ordinates with $y = 0, x > 0$ the undisturbed free surface and $y + x\tan\beta = 0$ the sloping beach. Notice the use in this section of k for the longshore wavenumber and, rather more perversely, the distinctly non-Manchester notation of y vertically upwards!

Ursell writes down the following edge wave solution:

$$
\begin{aligned}
\phi(x,y) \;=\;\; & e^{-kx\cos\beta + ky\sin\beta} \\
+ & \sum_{r=1}^{m} A_{rm}\{e^{-kx\cos(2r-1)\beta - ky\sin(2r-1)\beta} \\
+ & e^{-kx\cos(2r+1)\beta + ky\sin(2r+1)\beta}\}
\end{aligned}
\tag{11.23}
$$

where

$$A_{rm} = (-1)^r \prod_{s=1}^{r} \frac{\tan(m-s+1)\beta}{\tan(m+s)\beta} \tag{11.24}$$

and (11.18) is satisfied. Notice that the first term is just the Stokes solution. Clearly (11.24) satisfies (11.19) and (11.22) but it is not immediately obvious that conditions (11.20), (11.21) are satisfied by (11.24), (11.24), and a little work is needed to confirm this.

It is even less obvious as to how Ursell arrived at this solution. Incidentally, such higher order edge wave modes are frequently detected by oceanographers for waves over sloping beaches or shelving regions; see, for example, LeBlond & Mysak (1978, p. 227). Greenspan (1970) has extended this solution to include the boundary condition

$$\phi_n + \alpha\phi = 0 \quad \text{on the beach} \tag{11.25}$$

where α is constant and he too gives no explanation of his derivation. Both Roseau (1958) and Whitham (1979) provided systematic constructive methods which lead to solutions which can be manipulated into the Ursell solution. Peters (1952) also considered the problem but was more concerned with the case $k < K$.

Here we shall sketch a constructive derivation of the Greenspan solution based on the Whitham method showing how it reduces to the Ursell solution when $\alpha = 0$.

The idea is to look for solutions of the form

$$\phi(x,y) = \frac{1}{2\pi i}\int_C \{f(\zeta)E_+(\zeta) + g(\zeta)E_-(\zeta)\}\,d\zeta \tag{11.26}$$

where

$$E_\pm = \exp\frac{1}{2}k\{(\zeta + \zeta^{-1})x \pm i(\zeta - \zeta^{-1})y\},$$

satisfying (11.19) and seek f, g, C to satisfy (11.20) – (11.22). It can be shown that conditions (11.20) and (11.21) are satisfied if f satisfies

$$f(\zeta) = \zeta^{-1}h(\zeta,\mu)h(\zeta e^{i\beta},\chi) \tag{11.27}$$

where

$$h(\zeta,\mu) = \frac{(\zeta - l)(\zeta + \bar{l})}{(\zeta + l)(\zeta - \bar{l})}h(w\zeta,\mu) \tag{11.28}$$

and

$$w = e^{2i\beta}, \quad l = e^{i\mu}, \quad K = k\sin\mu, \quad p = e^{i\chi}, \quad \alpha = k\sin\chi.$$

The functional difference equation (11.28) can be solved by inspection if $\beta = \pi/2N$ or $w^N = -1$, since repeated application of (11.28) produces a relation between $h(\zeta,\mu)$ and $h(-\zeta,\mu)$ which is easily solved. Then from (11.27) we find

$$f(\zeta) = \zeta^{-1}\prod_{r=1}^{N}\frac{(\zeta - \bar{l}w^r)(\zeta - \bar{p}e^{-i\beta}w^r)}{(\zeta - lw^r)(\zeta - pe^{-i\beta}w^r)}$$

with a similar form for $g(\zeta)$.

Now an examination of the exponential terms in the integrands in (11.26) shows that f, g must be free of poles in the region $|\arg\zeta| < \frac{\pi}{2} + \beta$. Inspection of $f(\zeta)$ shows that this can only be achieved if

$$\bar{l}w^m = pe^{-i\beta} \quad \text{for some integer} \quad m \quad \text{or} \quad \mu = (2m+1)\beta - \chi$$

whence

$$f(\zeta) = \zeta^{-1}\frac{\zeta - l}{\zeta + l}\prod_{s=1}^{m}\frac{(\zeta - lw^{-s})(\zeta - \bar{l}w^s)}{(\zeta + lw^{-s})(\zeta + \bar{l}w^s)} \tag{11.29}$$

after re-arrangement. But now (11.29) is independent of N and indeed can be shown to satisfy (11.27) for *any* β, not just of the form $\pi/2N$. The expression for $g(\zeta)$ follows similarly and if C is chosen to enclose all the poles of $f(\zeta)$ except at $\zeta = 0$ the resulting form for ϕ reduces to an invigorating exercise in calculating residues. The result is

$$\begin{aligned}
\phi(x,y) &= e^{-kx\cos(\beta-\chi)+ky\sin(\beta-\chi)}\\
&+ \sum_{r=1}^{m}A_{rm}\left\{e^{-kx\cos\{(2r-1)\beta-\chi\}-ky\sin\{(2r-1)\beta-\chi\}}\right.\\
&+ \left. C_{rm}e^{-kx\cos\{(2r+1)\beta-\chi\}+ky\sin\{(2r+1)\beta-\chi\}}\right\}
\end{aligned}$$

where $\qquad K = k\sin\{(2m+1)\beta - \chi\}, \qquad \alpha = k\sin\chi,$

$$A_{rm} = (-1)^r \prod_{s=1}^{r} \frac{\tan(m-s+1)\beta}{\tan\{(m+s)\beta - \chi\}} \prod_{s=1}^{r-1} \frac{\tan(s\beta - \chi)}{\tan s\beta},$$

$$C_{rm} = \tan(r\beta - \chi)/\tan r\beta.$$

This is the Greenspan solution, reducing to the Ursell solution (11.24) as $\alpha \to 0$. Fuller details can be found in Evans (1989). Subsequently Packham (1989) has shown that the same results can be derived in a neater fashion, without any prior assumption as to the particular form of β, by using ideas of Williams (1959) for wedge problems.

In the light of the complexities of the above solution the question remains as to how Ursell knew that a solution in the form of sums of exponential, would give rise to a solution. Recent correspondence with Odulo (1989) has shed some light on this.

11.3   Submerged bodies

One of the most astonishing results of linear water wave theory is that suggested by Dean (1948) and first proved rigorously by Ursell (1950).

Consider, in two dimensions a sinusoidal wave-train incident upon a fixed rigid totally submerged circular cylinder in infinitely deep water. There are two independent dimensionless parameters in this problem, Ka, and Kf, where $K = \omega^2/g$ and ω is the incident wave frequency, a the cylinder radius, and f the depth of submergence of its centre. In the vertical barrier problem the reflection coefficient R has an explicit expression in terms of the single parameter $\mu = Ka$ given by (11.16). Despite being a function of two parameters, R for the submerged circular cylinder also has an explicit representation; it is precisely zero for all Ka, Kf. It follows from energy arguments that $|T| = 1$, the incident wave merely experiencing a phase shift as it passes over the cylinder.

The phenomenon of zero reflection is not new. It occurs at an infinity of discrete values of Ka for the scattering of waves by two barriers (Evans & Morris, 1972b) and in all situations involving multiple obstacles, but only for the submerged cylinder in deep water does it vanish identically for all Ka, Kf.

The Newman relations (Newman, 1975) provide a relation between radiation and scattering problems. In particular for a symmetric body making unit heave or sway oscillations, so that

$$\phi_H \sim H\exp(iK|x| - Ky), \qquad \text{as } |x| \to \infty \text{ in heave, and}$$
$$\phi_S \sim S\,\text{sgn}x\,\exp(iK|x| - Ky), \qquad \text{as } |x| \to \infty \text{ in sway,}$$

we have

$$R + T = \frac{-H}{\overline{H}} = -e^{2i\theta_H}$$

$$R - T = \frac{-S}{\overline{S}} = -e^{2i\theta_S}$$

where θ_H, θ_S, are the phases of H, S respectively. It follows that for the submerged circular cylinder, since $R = 0$,

$$\theta_S = \theta_H \pm \frac{\pi}{2}.$$

In fact Ogilvie (1963) showed that

$$S = iH \tag{11.30}$$

which result leads to further remarkable facts about the submerged cylinder.

Suppose the centre of the cylinder is forced to rotate in small circular clockwise motions, so that the motion of its centre (x_c, y_c) is

$$x_c = \delta \cos \omega t, \quad y_c - f = \delta \sin \omega t.$$

Then the velocity potential Φ satisfies, on linear theory,

$$\begin{aligned}
\frac{\partial \Phi}{\partial n} &= \dot{x}_c \sin \theta + \dot{y}_c \cos \theta = \delta \omega \cos (\omega t + \theta) \\
&= \delta \omega \operatorname{Re} \left\{ \frac{\partial \phi}{\partial n} e^{-i\omega t} \right\}
\end{aligned}$$

where

$$\phi = \phi_H - i\phi_S$$

and

$$\frac{\partial \phi}{\partial n} = \cos \theta - i \sin \theta = e^{-i\theta}.$$

It follows that

$$\phi(x, y) \sim (H - iS)e^{iKx - Ky}, \quad x \to +\infty$$

and

$$\begin{aligned}
\phi(x, y) &\sim (H + iS)e^{-iKx - Ky}, \quad x \to -\infty \\
&= 0
\end{aligned}$$

from (11.30). Thus there are no waves radiated to $-\infty$ by the clockwise circular motion of the cylinder. Time reversal now shows that any given wave-train from $x = +\infty$ can be absorbed by a suitable anti-clockwise circular motion of the cylinder of appropriate amplitude and phase. If the resistance and damping forces applied to the cylinder can be chosen correctly we have the capability of absorbing all the incident wave energy

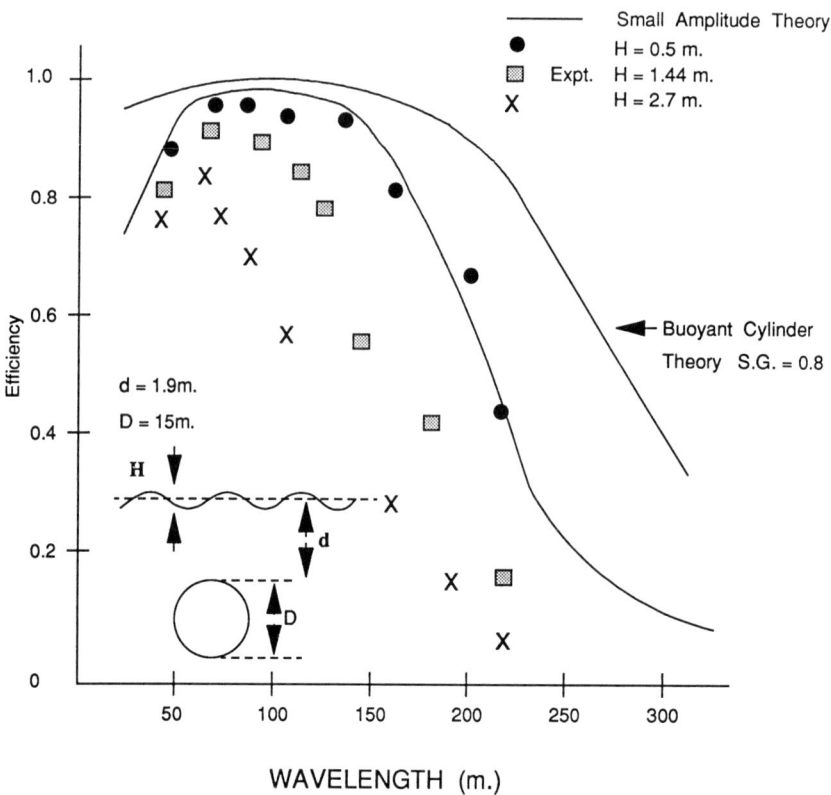

Figure 11.1: Efficiency of Bristol Cylinder Wave-Energy Device -
comparison of theory with experiment scaled up 150 times.

at that frequency (Evans 1976b). This idea formed the basis of the Bristol Submerged Cylinder wave-energy device first conceived in 1975 and developed in conjunction with Sir Robert McAlpine and Sons Ltd over the following years until the demise of the UK wave-energy programme in 1982. An impression of the efficiency of the cylinder device is given by figure 11.3 which compares theory with experimental results carried out at the University of Edinburgh in 1976 (Evans *et al.* 1979). The cylinder was constrained horizontally and vertically by electronically applied springs and dampers which could be varied at will to achieve tuning at the desired frequency. At full scale the device would be moored by cables or rodes making roughly 45° angles to the vertical with an integrated power take off system at the sea-bed moorings. Details are given in Clare *et al.* (1982).

A fixed submerged cylinder, a cylinder absorbing 100% of the incident wave-energy, or a neutrally buoyant cylinder responding to an incident wave train all share the property $R = 0$. In the second case $T = 0$ also. However it is possible to do just the opposite and utilize a submerged tethered buoyant cylinder as an effective moving breakwater. By securing the cylinder with vertical tethers from each end only small horizontal sway motions occur in plane incident waves and by a suitable choice of buoyancy and tether lengths, which in turn dictates the restoring force, wave cancellation can be achieved. In particular for a general symmetric body it can be shown (Evans & Linton 1989) that the relative amplitude of the wave past the body is

$$T_1 = T(C - \chi)/(C - i)$$

where R, T are the reflection and transmission coefficients for the fixed body, $\chi = -iR/T$, real, and $C = \{(M + I)\omega^2 - \lambda\}/B\omega$, where M, B are the added mass and damping coefficients for that body, I its inertia and $\lambda = (M' - I)g/l$ the restoring spring rate, where M' is the equivalent water mass. Figure 11.3 shows a comparison of theory with experiment for a circular cylinder in a narrow wave tank in Bristol. Notice that at a frequency of about 1.5 Hz theory predicts total wave cancellation and this is confirmed by the experiments. Over a range of frequencies corresponding to incident wavelengths between 0.4m and 1m, the transmission coefficient is below 50%. The idea of wave cancellation is of course not limited to submerged cylinders. However calculations on a totally-submerged vertical plate hinged at the sea-bed (Evans & Linton 1990) suggest that the equivalent submerged cylinder performs better in practice.

One final intriguing result is the following. Suppose we have a pulsating source in the presence of the fixed cylinder. If the source is sufficiently remote, the local field of the source will not influence the cylinder which will only experience a sinusoidal wave-train, which, in deep water, will be unaffected in amplitude by the cylinder since $R = 0$.

Thus the wave amplitude at either infinity due to the source will be

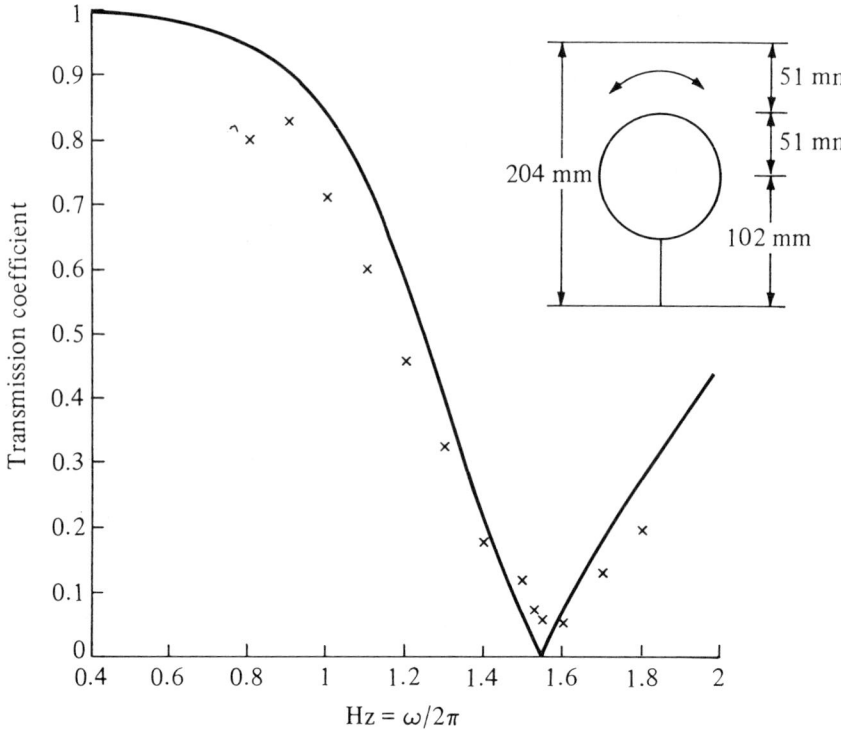

Figure 11.2: Transmission coefficient for submerged buoyant swaying cylinder – comparison of theory with experiment.

the same. The question can be asked, how close to the cylinder must the source be before the local field becomes important and this result ceases to be true? The answer, provided by a simple application of Green's theorem in a manner similar to that used by Newman (1975) in proving the Newman relations is that no matter how close the source is, even if it is on the cylinder, the radiated wave amplitude at each infinity is the same (Evans 1984). It has recently been shown (McIver & McIver 1990) that the result $R = 0$ is true to second-order in the wave amplitude also.

The existence of edge waves over a sloping beach has already been discussed. Such local solutions also exist when there is no obvious edge. In particular Ursell (1951), exploiting the geometric simplicity of the submerged circular cylinder showed that these trapped modes can travel along the top generator of a submerged horizontal circular cylinder without change, whilst decaying to zero in a horizontal direction perpendicular to the generators. He constructed a set of multipole potentials each satisfying all conditions except that of no flow through the cylinder.

By allowing the radius of the cylinder to become small whilst simultaneously approaching from below the cut-off frequency beyond which a continuous range of frequencies describing incident waves from infinity are possible, he was able to show that the resulting homogeneous infinite system of equations had a solution. Subsequent computations (McIver & Evans 1985) confirmed numerically the existence of other trapped modes for a submerged circular cylinder of arbitrary radius. Ursell (1987) has since demonstrated that a trapped mode always exists over a submerged cylinder of any cross-section, confirming the results of Jones (1953) by a simpler method. Prompted by a comment in this paper Callan (1990) has shown that trapped modes exist in the vicinity of a submerged sphere in a channel. Further results were reported at the 5th Workshop on Water Waves and Floating Bodies by Bonnet & Joly (1990).

I should like to conclude this review paper by describing some recent results on trapped modes in channels or equivalently, acoustic waveguides.

Fluid occupies an open channel bounded by $|y| = d$, $z = h$, with free surface $z = 0$, $|y| \leq d$, $-\infty < x < \infty$. A rectangular block is immersed in the channel throughout its depth and intersects the free surface at $|x| = a$, $|y| = b < d$.

Suppose the block is given an impulsive motion in the y direction, which is antisymmetric in y. The resulting fluid motion will be described by a superposition of all possible wave frequencies ω in a Fourier integral representation.

Now for motions odd in y, it can be shown by separation of variables that no wave propagation away from the block and down the channel is possible if $k < \pi/(2d)$ where k satisfies

$$\omega^2 = gk \tanh kh. \tag{11.31}$$

If such a frequency below the first cut-off exists in the presence of the block, then the ultimate fluid motion will consist of an undamped local oscillation at that frequency, all other modes having radiated their energy down the channel.

It has been shown numerically by Evans & Linton (1990) that such frequencies do indeed exist for all sizes of block and that the number of such frequencies increases as the horizontal dimension a increases. Such antisymmetric trapped modes are not confined to rectangular blocks and recently Callan et al. (1990) and McIver (1990) have shown their existence for small circular or general vertical cylinders in channels. Results for the block are given in figure 11.3. These modes can equally-well be regarded as trapped antisymmetric acoustic modes where now $k = \omega/c$ and c is the velocity of sound. It is believed that the trapped modes described here are new and may have interesting consequences. For example it can be shown that they also exist in a channel with a symmetric rectangular indentation in each wall showing the existence of seiche motions in unbounded regions.

11.4   Conclusion

Each of the problems described here has its origin in a notable paper by Fritz Ursell, whether it be on vertical barriers, sloping beaches,

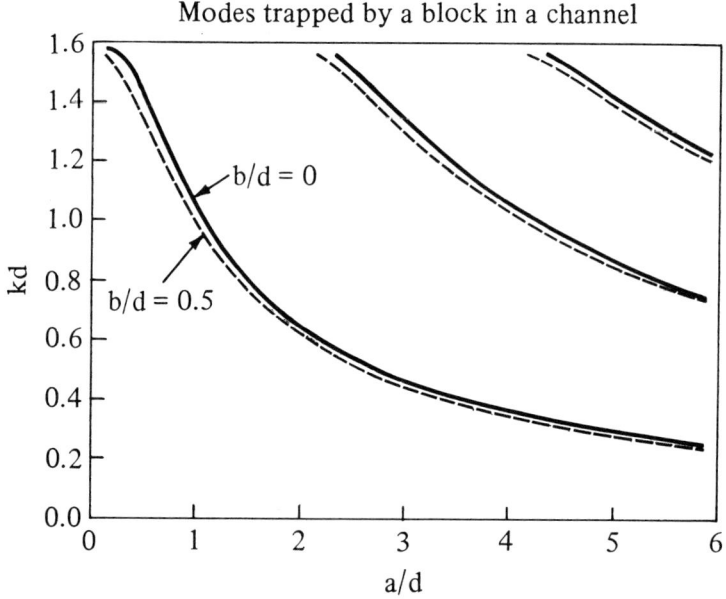

Figure 11.3: Trapped modes due to a rectangular block in a channel, as a function of block geometry.

or submerged circular cylinders. The exception is the last problem on trapped waves. That had its origin in a chance remark he made during a visit to Bristol in February 1990.

11.5 References

Bonnet, A.S & Joly, P. 1990 Mathematical and numerical study of trapped waves. Proc 5th Int. Workshop on Water Waves & Floating Bodies, 24–27.

Callan, M. 1990 Private communication.

Callan, M., Linton, C.M., Evans, D.V. 1990 To be submitted.

Clare, R., Evans, D.V. & Shaw, T.L. 1982 Harnessing sea wave energy by a submerged cylinder device.

Dean, W.R. 1945 On the reflexion of surface waves by a submerged plane barrier. *Proc. Camb. Phil. Soc.* **41**, 231–238.

Dean, W.R. 1948 On the reflection of surface waves by a submerged circular cylinder. *Proc. Camb. Phil. Soc.* **44**, 483–491.

Evans, D.V. 1976a A note on the waves produced by the small oscillations of a partially immersed vertical plate. *J. Inst. Maths. Applics.* **17**, 135–140.

Evans, D.V. 1976b A theory for wave power absorption by oscillating bodies. *J. Fluid Mech.* **77**, 1–25.

Evans, D.V. 1978 The oscillating water column wave-energy device. *J. Inst. Maths. Applics.* **22**, 423–433.

Evans, D.V. 1984 A note on the transparency of a submerged circular cylinder to the waves radiated by a pulsating line source. *IMA J. Appl. Maths.* **33**, 105–107.

Evans, D.V. 1989 Edge waves over a sloping beach. *Quart J. Mech. Appl. Maths.* **42**, 132–142.

Evans, D.V., Jeffrey, D.C., Salter, S.H. & Taylor, J.R.M. 1979 Submerged cylinder wave energy device: theory and experiment. *Appl. Ocean Res.* **1**, 3–12.

Evans, D.V. & Linton, C.M. 1989 Active devices for the reduction of wave intensity. *Appl. Ocean. Res.* **11**, 26–32.

Evans, D.V. & Linton, C.M. 1990 Submerged floating breakwaters. *Proc. 9th Int. Conf. Offshore Mech. Arctic Eng.*, Houston, **1**, part B, 279–286.

Evans, D.V. & Linton, C.M. 1990 Trapped modes in open channels. To be published in *J. Fluid Mech.*

Evans, D.V. & Morris, C.A.N. 1972a The effect of a fixed vertical barrier on obliquely incident surface waves in deep water. *J. Inst. Maths. Applics.* **9**, 198–204.

Evans, D.V. & Morris, C.A.N. 1972b Complementary approximations to the solution of a problem in water waves. *J. Inst. Maths. Applics.* **10**, 1–9.

Greenspan, H.P. 1970 A note on edge waves in a stratified fluid. *Stud. Appl. Maths.* **44**, 381–388.

Havelock, T.H. 1929 Forced surface-waves on water. *Phil. Mag.* **8**, 569–576.

Jones, D.S. 1953 The eigenvalues of $\nabla^2 u + \lambda u = 0$ when the boundary conditions are given in semi-infinite domains. *Proc. Camb. Phil. Soc.* **49**, 668–684.

Jones, D.S. 1964 *The theory of electromagnetism*. Oxford: Pergamon.

LeBlond, P.H. & Mysak, L.A. 1978 *Waves in the ocean*. New York: Elsevier.

Lewin, M. 1963 The effect of vertical barriers on progressing waves. *J. Math. Phys.* **42**, 287–300.

McIver, P. 1990 Private communication.

McIver, P. & McIver, M. 1990 Private communication.

McIver, P. & Evans, D.V. 1985 The trapping of surface waves above a submerged horizontal cylinder. *J. Fluid. Mech.* **151**, 243–255.

Mei, C.C. 1966 Radiation and scattering of transient gravity waves by vertical plates. *Quart. J. Mech. Appl. Math.* **19**, 417–440.

Newman, J.N. 1974 Interaction of water waves with two closely-spaced obstacles. *J. Fluid. Mech.* **66**, 97–106.

Newman, J.N. 1975 Interaction of waves with two-dimensional obstacles: a relation between the radiation and scattering problem. *J. Fluid Mech.* **71**, 273–282.

Odulo, A.B. 1990 Private communication.

Ogilvie, T.F. 1963 First- and second-order force on a cylinder submerged under a free surface. *J. Fluid Mech.* **16**, 451–472.

Packham, B.A. 1989 A note on generalised edge waves over a sloping beach. *Quart. J. Mech. Appl. Math.* **42**, 441–446.

Peters, A.S. 1952 Water waves over sloping beaches and the solution of a mixed boundary value problem for $\Delta^2\phi - k^2\phi = 0$ in a sector. *Comm. Pure Appl. Maths.* **5**, 87–108.

Porter, D. 1974 The radiation and scattering of surface waves by vertical barriers. *J. Fluid Mech.* **63**, 625–634.

Roseau, M. 1958 Short waves parallel to a shore over a sloping beach. *Comm. Pure Appl. Maths.* **6**, 433–493.

Stokes, G.G. 1846 Report on recent researches in hydrodynamics. *Brit. Assoc. Report.*

Ursell, F. 1947 The effect of a fixed vertical barrier on surface waves in deep water. *Proc. Camb. Phil. Soc.* **43**, 374–382.

Ursell, F. 1948 On the waves due to the rolling of a ship. *Quart. J. Mech. Appl. Math.* **1**, 246–252.

Ursell, F. 1950 Surface waves on deep water in the presence of a submerged circular cylinder I, II. *Proc. Camb. Phil. Soc.* **46**, 141–152, 153–163.

Ursell, F. 1951 Trapping modes in the theory of surface waves. *Proc. Camb. Phil. Soc.* **47**, 347–358.

Ursell, F. 1952 Edge waves over a sloping beach. *Proc. Roy. Soc. Lond. A* **214**, 79–97.

Ursell, F. 1987 Mathematical aspects of trapping modes in the theory of surface waves. *J. Fluid Mech.* **183**, 421–437.

Whitham, G.B. 1979 *Lectures on wave propagation.* New York: Springer.

Williams, W.E. 1959 Diffraction of an E-polarised plane wave by an imperfectly conducting wedge. *Proc. Roy. Soc. Lond. A* **252**, 376–393.

12

Some unsolved and unfinished problems in the theory of waves

F. Ursell
University of Manchester

> *The problems that I wish to discuss in the present paper are problems on which I have worked at various times during my career. In every case some progress has been made but every step forward has also revealed new problems which deserve further study. The selection is somewhat arbitrary, I have never yet solved a problem completely.*

12.1   The uniqueness problem for linear water waves

This is one of the most fundamental problems in the linear theory of water waves. The problem is shown in Figure 12.1, for either two or three dimensions. The parameter $K = \omega^2/g$ is positive. We wish to know whether the homogeneous equations and boundary conditions shown in Figure 12.1 imply that ϕ vanishes identically. This has been established for certain restricted classes of surfaces S in three dimensions (or curves in two dimensions) but in general the answer is unknown. The uniqueness problem can be rephrased: If K is regarded as a spectral parameter the spectrum is continuous for $0 < K < \infty$; we wish to know whether there are

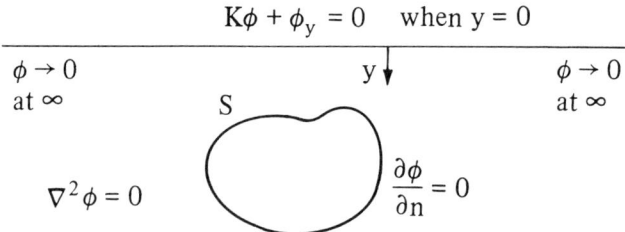

Figure 12.1: The uniqueness problem for linear water waves

discrete values of K (corresponding to modes of finite energy) embedded in this continuous spectrum. References: W.6[1] and W.33.

12.2 Problems relating to the heaving motion of a circular cylinder in the surface of a fluid (moderate and long wavelengths)

As is well known, this problem can be solved by integral equations of the second kind, the standard method of potential theory (see W.12 and §12.3 below). An alternative method, the method of multipoles, was given in 1949 (see also W.34). The potential is written as the sum of a wave source and of wavefree potentials

$$\phi(r\sin\theta, r\cos\theta) = C\Phi_0(Kr, \theta)$$
$$+ \ C\sum_{m=1}^{\infty} p_{2m}(Ka)a^{2m}\Phi_m(r,\theta),$$

where

$$\Phi_0(Kr,\theta) = \fint_0^\infty e^{-kr\cos\theta}\cos(kr\sin\theta)\,\frac{dk}{k-K}$$

and

$$\Phi_m = \frac{\cos 2m\theta}{r^{2m}} + \frac{K}{2m-1}\frac{\cos(2m-1)\theta}{r^{2m-1}}.$$

The boundary condition on the semi-circle

$$\left\langle\frac{\partial\phi}{\partial r}\right\rangle = V_0\cos\theta \quad \text{when} \quad r = a, \quad 0 \le \theta \le \frac{1}{2}\pi \tag{12.1}$$

is satisfied if

$$\left\langle a\frac{\partial\Phi_0}{\partial r}\right\rangle_{r=a} = \frac{aV_0}{C}\cos\theta$$
$$+ \ \sum_{m=1}^{\infty} 2m\cdot p_{2m}\left(\cos 2m\theta + \frac{Ka}{2m}\cos(2m-1)\theta\right),$$

for $0 \le \theta \le \frac{1}{2}\pi$; in other words, (12.1) is satisfied if the known function on the left can be expanded in terms of the infinite set of non-orthogonal functions on the right. This was the method adopted in W.3; more precisely, the expansion was integrated once with respect to θ, and the coefficients were fitted by least squares. In later work (W.26a) the expansion was multiplied by $\cos 2s\theta$ ($s = 0, 1, 2, \ldots$) and integrated over $(0, \frac{1}{2}\pi)$, and an infinite system

$$c_{2s}(Ka) = 2s \cdot p_{2s} + Ka\sum_{m=1}^{\infty}(2m\cdot p_{2m})b_{sm}, \quad s = 1,2,3,\ldots$$

[1]Citations are to papers listed in § 12.8

was thus obtained, where the coefficients b_{sm} are simple explicit rational functions of s and m while the coefficients $c_{2m}(Ka)$ are given by infinite series (converging rapidly when Ka is not large). (See W.26a.) The virtual mass and damping then involve simple linear sums over the unknowns $p_{2m}(Ka)$.

It can be shown that the unknowns $p_{2m}(Ka)$ decrease like $1/m^3$, see W.26a, and it follows that the three series for the potential and for the velocity components converge on the semi-circle. There is a weak singularity in the corners $r = a$, $\theta = \pm\frac{1}{2}\pi$; if the form of the singularity were known we would be able to subtract it out and greatly increase the rate of convergence. To fix ideas, consider the reflection problem, then $\partial\phi/\partial r = 0$ when $r = a$, $-\frac{1}{2}\pi \leq \theta \leq \frac{1}{2}\pi$. As is well known (see W.23), by means of the free-surface condition the potential can be continued analytically into the whole plane exterior to $r = a$, with a cut along $\theta = \pm\pi$; also ϕ can be continued into the semi-circle by means of the boundary condition $\partial\phi/\partial r = 0$ from which it follows that $\phi(r, \theta) = \phi(a^2/r, \theta)$ when $\frac{1}{2}\pi \leq \theta \leq \frac{1}{2}\pi$. These two continuations do not match along the quadrants $r = a$, $\frac{1}{2}\pi \leq \theta \leq \pi$ and $r = a$, $-\pi \leq \theta \leq -\frac{1}{2}\pi$ which are therefore cuts originating in the two corners, but it does not seem possible to find simple relations for the discontinuities in ϕ and $\partial\phi/\partial r$ across these cuts. It is easy to see that these cuts can be moved, for example to the diameter $-a \leq x \leq a$, $y = 0$, but this does not seem to help.

Problem 1 *In practice the infinite system is truncated to form an $M \times M$ system; what is the truncation error in the virtual mass and damping? If we could find an asymptotic expression for this error we could use it to extrapolate to $M = \infty$.*

Problem 2 *Can the virtual mass and damping be found explicitly as functions of Ka? We know (see W.24) that the complex force-coefficient is of the form*

$$\Lambda(Ka) = \frac{(\log Ka - \pi i)A_1(Ka) + A_2(Ka)}{(\log Ka - \pi i)A_3(Ka) + A_4(Ka)},$$

where $A_1(Ka)$, $A_2(Ka)$, $A_3(Ka)$ and $A_4(Ka)$ are entire functions of Ka which are real for real Ka. Can we find these functions? (Obviously there can be an arbitrary common factor.) Is there perhaps an explicit solution for the heaving problem?

Problem 3 *The corner singularity.*

12.3   Short-wave asymptotics for the heaving circular cylinder

The method of §12.2 above is not effective for large values of $Ka = N$, say. The same problem can also be formulated as an integral equation of the second kind for the values $\phi(\alpha, N)$ of the potential on the semi-circle:

$$\pi\phi(\alpha, N) \;+\; \int \phi(\theta, N)\frac{\partial}{\partial n}G(\theta, \alpha, N)\,a\,d\theta$$

$$=\; \int \frac{\partial\phi}{\partial n}(\theta)G(\theta, \alpha, N)\,a\,d\theta,$$

where $G(\theta, \alpha, N)$ and $\partial G(\theta, \alpha, N)/\partial n$ are derived from a fundamental solution (source potential), see W.12. When G is the simple wave-source potential this equation has irregular values which can be removed by putting a suitable additional wave source at the centre of the circle. (It is not difficult to see physically why this should work, see W.12 and W.30.) The kernel of the integral equation then tends to 0 when $N \to \infty$, the integral equation can be solved iteratively for sufficiently large values of N, and the process is convergent. The leading terms in the virtual mass and damping (wavemaking) are found to be $1 - (4/3\pi N)$ and $4/N$ respectively. The transient problem for the heaving circular cylinder can be solved by Laplace transforms (see W.24), and the solution involves the complex force coefficient $\Lambda(\omega(a/g)^{\frac{1}{2}})$ for complex frequencies ω. This is defined in the whole ω-plane cut from $\omega = 0$ to $\omega = -i\infty$ (see Problem 2 above), and from the integral equation the asymptotic relation

$$\Lambda(\omega(a/g)^{\frac{1}{2}}) \sim 1 - (4g/3\pi a\omega^2) + \cdots \quad \text{when} \quad |\,\omega(a/g)^{\frac{1}{2}}\,| \to \infty$$

is easily shown to be valid in the upper half ω-plane but its validity is also needed in part of the lower half ω-plane. By modifying the integral equation it has been shown that the relation is still valid when $-\pi/8 < \arg(\omega(a/g)^{\frac{1}{2}}) < 9\pi/8$, but it is clear that the mathematical argument is capable of improvement.

Problem 4 *Higher terms for the virtual mass and wavemaking can be found (in principle the complete expansion can be found) and involve terms like* $(\log N)^{-p}N^{-q}$. *The decrease is slow, and the two series (though convergent) are not numerically effective. Is there a more effective process, perhaps associated with a different kernel for the integral equation?*

Problem 5 *The same for the more difficult transmission problem, see W.18. (Amplitude transmission coefficient* $\sim 2/\pi N^4$.)

Problem 6 *As has just been noted, the asymptotic relation for the complex force-coefficient is known to be valid for complex frequencies in the sector from* $-\pi/8$ *to* $9\pi/8$. *In fact computations by Maskell have shown that the region of validity includes the sector from* $-\pi/4$ *to* $5\pi/4$, *but this has not yet been proved mathematically.*

12.4   Problems relating to wave resistance

In the mathematical theory of wave-resistance a (thin) ship moving with constant forward speed is represented by a distribution of Kelvin sources over its mid-plane. The free-surface boundary condition is

$$U^2\phi_{xx} - g\phi_y = 0 \quad \text{on the mean free surface } y = 0.$$

The potential of a submerged wave source can be expressed in several different ways as the sum of a single integral and a double integral, where only the single integral contributes to the wave pattern at a distance from the disturbance. (See W.16, W.36 and W.37.) Mathematically it is sufficient to consider a source in the free surface. It is known (see W.16) that the single integral is singular along the entire track of the free surface.

Problem 7 *Expansion of the near field of the Kelvin source. Although the expansion of the single integral near the source has recently been obtained (W.37), the form of the expansion of the complete potential near the source is still unknown.*

12.5   Problems relating to trapping modes

Consider the two-dimensional uniqueness problem for water waves when Laplace's equation

$$\phi_{xx} + \phi_{yy} = 0$$

is replaced by the modified Helmholtz equation

$$\phi_{xx} + \phi_{yy} - k^2\phi = 0;$$

the continuous spectrum then extends from $K = k$ to $K = \infty$. It was shown by Stokes that for a sloping beach there is a discrete point in the spectrum between $K = 0$ and $K = k$, and for beaches of small angle this is only one of several modes (W.9). The corresponding characteristic modes of finite energy are known as trapping modes or trapped modes. The existence of a trapping mode for fluid bounded internally by a submerged small circle was shown in W.7 and was extended to a wide class of curves by D.S. Jones. A simpler proof was given more recently in W.35. Some of the arguments in W.35 are incomplete.

Problem 8 *It has not yet been shown rigorously that the number of trapping modes increases indefinitely as the submerged body approaches the free surface.*

Problem 9 *It was also suggested but not proved that trapping modes may be possible near a submerged sphere in a canal of finite width.*

12.6 Problems relating to integrals with a large parameter

The Method of Steepest Descents is the standard method for evaluating contour integrals containing a large parameter, but it breaks down in its simplest form when two or more saddle points are nearly coincident, as happens near the critical lines in the Kelvin ship-wave pattern. A solution for the problem of two nearly coincident saddle points was given by Chester, Friedman and Ursell (M.1) and later applied to the Kelvin pattern (W.16); a solution for an arbitrary number of nearly coincident saddle points was given in M.5. These papers establish the validity of the relevant asymptotic expansions in domains which do not shrink to zero as the large parameter tends to infinity, nevertheless many problems remain.

Problem 10 *Successive terms in the expansions are obtained by means of the Bleistein Sequence, a convergent procedure, but there is as yet no method for obtaining efficient numerical bounds, particularly in unbounded domains. In this respect asymptotic expansions obtained from integrals are still inferior to expansions obtained from ordinary differential equations.*

12.7 Conclusions

There are many more unsolved and unfinished problems which I could add to the preceding list. They are of all degrees of difficulty and I hope that I am contributing to their solution by bringing them to the attention of a wider public.

12.8 Publications by F. Ursell

12.8.1 Water waves

W.1 The effect of a fixed vertical barrier on surface waves in deep water. *Proc. Camb. Phil. Soc.* **43** (1947) 374–382.

W.2 On the waves due to the rolling of a ship. *Quart. J. Mech. Appl. Math.* **1** (1948) 246–252.

W.3 On the heaving motion of a circular cylinder on the surface of a fluid. *Quart. J. Mech. Appl. Math.* **2** (1949) 218–231.

W.4 On the rolling motion of cylinders in the surface of a fluid. *Quart. J. Mech. Appl. Math.* **2** (1949) 335–353.

W.5 Surface waves on deep water in the presence of a submerged circular cylinder. I. *Proc. Camb. Phil. Soc.* **46** (1950) 141–152.

W.6 Surface waves on deep water in the presence of a submerged circular cylinder. II. *Proc. Camb. Phil. Soc.* **46** (1950) 153–158.

W.7 Trapping modes in the theory of surface waves. *Proc. Camb. Phil. Soc.* **47** (1951) 347–358.

W.8 Discrete and continuous spectra in the theory of gravity waves. In *Gravity waves*, Proceedings of symposium held at National Bureau of Standards, Washington, June 1951, N.B.S. Circular 521, pp. 1–5.

W.9 Edge waves on a sloping beach. *Proc. Roy. Soc.* A**214** (1952) 79–97.

W.10 Mass transport in gravity waves. *Proc. Camb. Phil. Soc.* **49** (1953) 145–150.

W.11 The long-wave paradox in the theory of gravity waves. *Proc. Camb. Phil. Soc.* **49** (1953) 685–694.

W.12 Short surface waves due to an oscillating immersed body. *Proc. Roy. Soc.* A**220** (1953) 90–103.

W.13 Water waves generated by oscillating bodies. *Quart. J. Mech. Appl. Math.* **7** (1954) 427–437.

W.14 (with T.B. Benjamin) The stability of the plane free surface of a liquid in vertical periodic motion. *Proc. Roy. Soc.* A**225** (1954) 505–515.

W.14a On the virtual mass and damping of ships at zero speed ahead. *Proc. Symp. Behaviour of ships in a seaway*, Wageningen 1959, pp. 374–387.

W.15 (with R.G. Dean & Y.S. Yu) Forced small-amplitude water waves: comparison of theory and experiment. *J. Fluid Mech.* **7** (1959) 33–52.

W.16 On Kelvin's ship wave pattern. *J. Fluid Mech.* **8** (1960) 418–431.

W.17 Steady wave patterns on a non-uniform steady fluid flow. *J. Fluid Mech.* **9** (1960) 333–346.

W.18 The transmission of surface waves under surface obstacles. *Proc. Camb. Phil. Soc.* **57** (1961) 638–668.

W.19 (with Y.S. Yu) Surface waves generated by an oscillating circular cylinder on shallow water: theory and experiment. *J. Fluid Mech.* **11** (1961) 529–551.

W.20 Slender oscillating ships at zero forward speed. *J. Fluid Mech.* **14** (1962) 496–516.

W.21 The decay of the free motion of a floating body. *J. Fluid Mech.* **19** (1964) 305–319.

W.22 On head seas travelling along a horizontal cylinder. *J. Inst. Math. Applns.* **4** (1968) 414–427.

W.23 The expansion of water-wave potentials at great distances. *Proc. Camb. Phil. Soc.* **64** (1968) 811–826.

W.24 (with S.J. Maskell) The transient motion of a floating body. *J. Fluid Mech.* **44** (1970) 303–313.

W.25 (with W.E. Bolton) The wave force on an infinitely long circular cylinder in an oblique sea. *J. Fluid Mech.* **57** (1973) 241–256.

W.26 Short surface waves in a canal: dependence of frequency on curvature. *J. Fluid Mech.* **63** (1974) 177–181.

W.26a A problem in the theory of water waves. In *Numerical solution of integral equations* (ed. L.M. Delves & J. Walsh), pp. 291–299. Oxford University Press 1974.

W.27 The refraction of head seas by a long ship. *J. Fluid Mech.* **67** (1975) 689–703.

W.28 On the virtual-mass and damping coefficients in water of finite depth. *J. Fluid Mech.* (1976) 17–28.

W.29 The refraction of head seas by a long ship. Part 2. Waves of long wavelength. *J. Fluid Mech.* **82** (1977) 643–657.

W.30 Irregular frequencies and the motion of floating bodies. *J. Fluid Mech.* **105** (1981) 143–156.

W.31 Mathematical notes on the two-dimensional Kelvin-Neumann problem. *Proc. 13th Symp. on Naval Hydrodynamics*, Tokyo 1980, pp. 245–255. (Tokyo: Shipbuilding Research Association of Japan 1981).

W.32 Mathematical note on the fundamental solution (Kelvin source) in ship hydrodynamics. *IMA J. Appl. Math.* **32** (1984) 335–351.

W.33 (with M.J. Simon) Uniqueness in linearized two-dimensional water-wave problems. *J. Fluid Mech.* **148** (1984) 137–154.

W.34 Mathematical observations on the method of multipoles. *Schiffstechnik* **33** (1986) 113–128.

W.35 Mathematical aspects of trapping modes in the theory of surface waves. *J. Fluid Mech.* **183** (1987) 421–437.

W.36 On the theory of the Kelvin ship-wave source: asymptotic expansion of an integral. *Proc. Roy. Soc.* A**418** (1988) 81–93.

W.37 On the theory of the Kelvin ship-wave source: the near-field convergent expansion of an integral. *Proc. Roy. Soc.* A**428** (1990) 15–26.

W.38 Some mathematical contributions to waves and ships. In *Dynamics of marine vehicles and structures in waves*, Proc. IUTAM Symposium, Brunel University, June 1990 (ed. W.G. Price). Elsevier, Amsterdam, to appear.

12.8.2 Aerodynamics

A.1 Notes on the linear theory of incompressible flow around symmetrical swept-back wings at zero lift. *Aeronautical Quarterly* **1** (1949) 101–122.

A.2 (with G.N. Ward) On some general theorems in the linearized theory of compressible flow. *Quart. J. Mech. Appl. Math.* **3** (1951) 326–348.

12.8.3 Acoustics

B.1 On the short-wave asymptotic theory of the wave equation $(\nabla^2 + k^2)\phi = 0$. *Proc. Camb. Phil. Soc.* **53** (1957) 115–133.

B.2 On the rigorous foundation of short-wave asymptotics. *Proc. Camb. Phil. Soc.* **62** (1966) 227–244.

B.3 Creeping modes in a shadow. *Proc. Camb. Phil. Soc.* **64** (1968) 171–191.

B.4 On the exterior problems of acoustics. *Proc. Camb. Phil. Soc.* **74** (1973) 117–125.

B.5 On the exterior problems of acoustics. II. *Math. Proc. Camb. Phil. Soc.* **84** (1978) 545–548.

12.8.4   Mathematical methods

M.1 (with C. Chester & B. Friedman) An extension of the method of steepest descents. *Proc. Camb. Phil. Soc.* **53** (1957) 599–611.

M.2 Integrals with a large parameter. The continuation of uniformly asymptotic expansions. *Proc. Camb. Phil. Soc.* **61** (1965) 113–128.

M.3 Integral equations with a rapidly oscillating kernel. *J. London Math. Soc.* **44** (1969) 449–459.

M.4 Integrals with a large parameter. Paths of descent and conformal mapping. *Proc. Camb. Phil. Soc.* **67** (1970) 371–381.

M.5 Integrals with a large parameter. Several nearly coincident saddle points. *Proc. Camb. Phil. Soc.* **72** (1972) 49–65.

M.5a Introduction to the theory of linear integral equations. In *Numerical solution of integral equations* (ed. L.M. Delves & J. Walsh), pp. 2–11. Oxford University Press 1974.

M.6 Integrals with a large parameter: a double complex integral with four nearly coincident saddle points. *Math. Proc. Camb. Phil. Soc.* **87** (1980) 249–273.

M.7 Integrals with a large parameter: Hilbert transforms. *Math. Proc. Camb. Phil. Soc.* **93** (1983) 141–149.

M.8 Integrals with a large parameter: Legendre functions of large degree and fixed order. *Math. Proc. Camb. Phil. Soc.* **95** (1984) 367–380.

M.9 Uniformly asymptotic expansions for an integral with a large and a small parameter. *Math. Proc. Camb. Phil. Soc.* **101** (1987) 349–362.

M.10 Integrals with a large parameter. A strong form of Watson's Lemma. In *Elasticity, mathematical methods and applications*, The Ian N. Sneddon 70th Birthday Volume (ed. G. Eason & R.W. Ogden), pp. 391–395. Ellis Horwood, Chichester 1990.

M.11 Integrals with a large parameter and the maximum-modulus principle. In *Asymptotic and computational analysis*, Conference in honour of Frank W.J. Olver's 65th birthday (ed. R. Wong), pp. 477–489. Marcel Dekker, New York 1990.

12.8.5   Oceanography

O.1 (with N.F. Barber) The response of a resonant system to a gliding tone. *Phil. Mag.*, Ser. 7 **39** (1948) 345–361.

O.2 (with N.F. Barber) The generation and propagation of ocean waves and swell. I. Wave periods and velocities. *Phil. Trans. Roy. Soc.* A**240** (1948) 527–560.

O.3 On the theoretical form of ocean swell on a rotating earth. *Monthly Notices of the Royal Astronomical Soc. Geophys. Suppl.* **6** (1950)

1–8.

O.4 On the application of harmonic analysis to ocean wave research. *Science* III (1950) 445–446.

O.5 (with M.S. Longuet-Higgins) Sea waves and microseisms. *Nature* **162** (1948) 700.

O.6 (with N.F. Barber, J. Darbyshire & M.J. Tucker) A frequency analyser used in the study of ocean waves. *Nature* **158** (1946) 329.

O.7 Wave generation by wind. In *Surveys in mechanics* (ed. G.K. Batchelor & R.M. Davies), pp. 216–249. Cambridge University Press 1956.

12.8.6 Biography

R.1 (with D.E. Cartwright) Joseph Proudman 1888–1975. *Biographical Memoirs of Fellows of the Royal Society* **22** (1976) 319–333.

12.9 Notes on the list of publications

In the list the publications are tabulated under subject headings. In these notes an attempt is made to indicate the logical sequence and mutual influence of the various contributions. The following major directions can be distinguished.

1. Ocean waves
2. Ship hydrodynamics
3. Mathematical theory of the linearized boundary value problems of water waves
4. Partial differential equations with a large parameter
5. Integrals with a large parameter
6. Experimental study of the predictions of linearized theory
7. Acoustics

No attempt will be made to review the general literature in these fields.

12.9.1 Ship hydrodynamics and linear theory of water waves

W.1 and W.2 These are my first papers on ship hydrodynamics and the mathematical study of boundary value problems in the theory of water waves. Consider a ship in waves; what are the forces on the ship? As is well-known, ship theory has to make drastic assumptions, and has had very limited success. Linearized theories have been developed by Havelock for the infinitely light ship (travelling pressure distribution) and for the infinitely thin ship (Michell ship) and the deeply submerged ship. Since then we have had the slender-ship approximation, see W.20. My own work is mostly concerned with the ship at zero forward speed (or at constant forward speed in a strip theory). In W.1 and W.2 the ship is replaced by its vertical mid-plane (fixed in W.1, oscillating in W.2) and an explicit solution is found. The solution is applied in W.2 to a problem of W. Froude and Havelock: to calculate the waves due to a rolling ship of known dimensions. Havelock had considered a submerged ship and found that the calculated waves are smaller than the observed waves. My calculation in W.2 gave very close agreement with observation, but nevertheless (as I pointed out) this is fortuitous: the ship's thickness cannot be neglected (see W.4). The solutions have been rediscovered by several authors. Sequel: W.4.

W.3 and W.4   These are concerned with the forces on a ship oscillating near a free surface. The velocity potential is expanded in terms of a *wave source* (or dipole) and *wave-free potentials*. The coefficients can be calculated from an infinite system of linear equations; the solution can be shown to exist. Virtual-mass and damping coefficients are computed in the range $0 \le Ka \le 4.5$, the first such computations in the literature. The virtual-mass coefficient is shown to be strongly frequency-dependent. In W.4 the method is extended to rolling: it is shown (among other applications) that certain ship-like sections generate no waves in rolling. Their rolling motion (without bilge keels) is thus undamped by waves. This has since been verified experimentally.

Coefficients of virtual mass and damping have since been computed for many cross-sections, essentially by this method. The method is not suited for high frequencies, $Ka > 4.5$. Sequels: W.12 and W.26a.

W.5 and W.6   This work was inspired by a remarkable result due to W.R. Dean: Consider water waves incident on a submerged circular cylinder. The reflection coefficient is a function of two independent parameters; Dean found it to be zero. His mathematical argument was incomplete but the result is correct, as is shown in W.5 where a rigorous solution by infinite determinants is given. There remains the problem of uniqueness, still unsolved in general but here demonstrated (in W.6) for the submerged circular cylinder. No significant progress has been made with the uniqueness problem since that date. Sequels: W.7 and W.8.

W.7 and W.8   What can be said about uniqueness when the equation of continuity $\nabla^2 \phi = 0$ of W.7 is replaced by $(\nabla^2 - k^2)\phi = 0$? A counter-example to uniqueness is then Stokes's edge wave on a sloping beach (1846). Can similar modes be found for the submerged circular cylinder, for example? It is shown in W.7 that the condition for this is the vanishing of a certain very complicated infinite determinant. A limiting procedure reduces the determinant to a simple form when certain parameters vanish. It follows by continuity that the determinant can vanish when these parameters are small and appropriately chosen.

These results were greatly extended by D.S. Jones using functional analysis. His results do not depend on small parameters. Sequels: W.9 and W.35.

W.9   An experimental and theoretical study of trapping modes on a sloping beach, see W.7. Edge waves were excited in a small tank, and the wave amplitude was measured as a function of frequency. At the frequency of the Stokes mode a resonance was expected and observed. A resonance was also observed at the cut-off frequency which however was found to depend only on the width of the tank near the wave-maker, not on the width far down the tank. (This is attributed to the action of viscosity at zero group velocity.) It was further shown theoretically that the Stokes mode is only one of a family of modes; the number of modes increases as the beach angle decreases. Measurements were made near 30°, the first critical angle where a resonance frequency coincides with the cut-off frequency. A very large amplitude was observed. The agreement between calculated and observed frequencies is close.

W.10   A simple geometrical argument for the existence of mass transport in periodic irrotational waves was given by Lord Rayleigh for deep water. A simple analytical treatment is given here and is shown to be applicable to finite depth and to solitary waves. (Similar results were obtained by Longuet-Higgins.)

W.11   Standard texts (such as Lamb's *Hydrodynamics*) show that there is an analogy between long water waves and acoustic waves; if such an analogy is valid, then long waves must ultimately break. But solitary waves are long and do not break. The paradox is resolved by a formal expansion involving *two* small parameters; their relative magnitude is seen to be important.

W.12   This is a sequel to W.3; the virtual-mass and damping calculations given there became ill-conditioned for $Ka > 4.5$. A different process is devised here for large Ka. The problem is now set up as an integral equation for the boundary value $\phi(\theta, Ka)$ of the potential on the semi-circle. It is expected that this will tend to a limit $\phi(\theta, \infty)$, but this is difficult to show since the kernel of the integral equation is found to tend to infinity near irregular frequencies. It is noted, however, that there are infinitely many integral equations for $\phi(\theta, Ka)$ depending on the choice of fundamental solution, and the kernels of some of these tend to limiting kernels. Asymptotic expressions for virtual mass and damping are rigorously determined in this way. The principal interest of this work is however theoretical. Before this the short-wave limit (wave optics to ray optics, wave mechanics to classical mechanics) had been studied rigorously only in a few cases where an explicit solution could be found. In the present paper there is no explicit solution, and nevertheless a rigorous discussion is possible. Sequels: W.13, W.14a, W.18, W.26 and B.1.

W.13   The results of W.12 are derived by plausible non-rigorous arguments. Earlier plausible arguments due to Prandtl and his students are shown to be incorrect.

W.14   Stability of the free surface of a liquid in vertical periodic motion. The theory is due to both authors, the experiment to T.B. Benjamin. Agreement is very close.

W.14a   In this paper the virtual mass and damping for a half-immersed circle are given for the first time over the whole range $0 \le Ka < \infty$; the results of W.12 are used to interpolate between $Ka = 4.5$ and $Ka = \infty$.

W.15 and W.19   Linearized irrotational theory of water waves has been extensively used for water wave calculations. How good is the agreement with experiment? Good agreement for group velocity was found in O.2, and for frequencies in W.9, and similar confirmation had been found by other workers, but there was little evidence for amplitudes. This is given in the present set of papers. The experimental difficulties arise from wave reflection at the ends of the wave tank, which is rarely less than 10% of the incident amplitude. In the present work the geometry is two-dimensional, and it is shown that the reflected wave can be analysed away; agreement of the measured and corrected amplitude with theoretical predictions is then very close in a certain range of wave steepnesses. The co-authors were two very good graduate students at the M.I.T. Hydrodynamics Laboratory (both now professors of civil engineering).

W.16   (Sequel to M.1.) The Kelvin ship wave pattern can be described with sufficient accuracy by a single integral

$$\int (1 + u^2) \exp \left[iN(\cos\theta - u\sin\theta)\sqrt{1 + u^2} \right] du,$$

as is well known. The problem reduces to an asymptotic study of the integral for large N. Kelvin himself applied his principle of stationary phase to find the wave pattern everywhere except near certain critical lines. Havelock found the amplitude *on* the critical lines, but the behaviour *near* the critical lines remained unsolved. This problem is a special case of a well-known problem in uniform asymptotics, studied by G.B. Airy (in his work on caustics), G.N. Watson (in his work on Bessel functions) and many others, but not adequately solved by them. The solution developed in M.1 is here applied to the Kelvin pattern. (Computations by H.P.F. Swinnerton-Dyer.)

W.17   This problem was suggested by Sir Geoffrey Taylor's work on water bells. A steady wave pattern is formed on a steady non-uniform flow by a point disturbance. A kinematic approximate theory is here given, which is an extension to two dimensions of the well-known kinematic theory of group velocity. The existence of a set of characteristic curves is established; if these are assumed to pass through the point disturbance, then the wave pattern can be constructed. Amplitude variations are not treated. It would be difficult to make this work rigorous, or to express it in terms of a small-parameter expansion.

W.18   This is a sequel to W.12. Short surface waves are incident on a fixed circular cylinder; what is the amplitude of the transmitted wave? The answer to this question contributes to an understanding of the mechanism of transmission under the cylinder. In theory the method of W.12 is applicable but unfortunately four iterations are required when the obvious formulation is used. The direct approach is therefore not practicable. It is explained how the transmission coefficient can nevertheless be obtained from simple rules, and it is shown mathematically that the result is correct. The calculation is somewhat formidable.

Considerable efforts have since been made to translate the mathematical arguments into physically plausible assumptions. Ingenious formal procedures have been devised by F.G. Leppington and by A.M.J. Davis.

W.20   As has already been explained (W.1 above), in ship hydrodynamics it is impossible to avoid drastic assumptions. At that time (1960) it was hoped that the slender ship (beam and draught much smaller than length and wavelength) would prove a good model. The present paper was followed by the work of E.O. Tuck, Joosen and many others. The application to ships was unfortunately disappointing but in its more elaborate forms (strip theory) it is now widely applied. Sequels: W.22 and W.25.

W.21   This is a sequel to W.14a. Since the virtual mass and damping are known for *all* frequencies, the transient motion of a floating body following an initial displacement or velocity can be determined in principle. Expressions are given in the form of Fourier integrals. The motion after a long time can be found from these provided that the virtual mass and damping can be continued into the lower half of the frequency plane (this is easy to show) and that they behave suitably in a sector adjoining the real frequency axis (this is difficult). An argument to prove the latter result is outlined, and it is inferred that the transient motion is ultimately non-oscillatory. The details of the argument have since been supplied by Crapper. Sequel: W.24.

W.22   The papers W.22, W.23 and W.25 were all motivated by the developing interest in strip theory. In this theory, each section of the ship is treated as if it were a portion of an infinitely long cylinder of the same cross-section. The calculations of virtual mass and damping (originated in W.3, W.4, W.12, W.13, W.14a and developed by many authors) are often applied but these calculations refer to beam incidence. In practice it is oblique incidence or head incidence that is relevant.

In W.22 head seas are considered; this is probably the most interesting case in practice. It is therefore unfortunate that head seas cannot travel along an infinite cylinder without change of form as is shown here. The progressive transformation of such waves is considered in W.27 and W.29.

W.23   The expansion of wave potentials in terms of wave sources, wave dipoles and wave-free potentials is still valid at oblique incidence. This result is proved here by arguments involving analytic continuation. Mainly for reference in W.22 and W.25.

W.24   Sequel to W.21. Numerical computation of the results of W.21. The Fourier integrals are slowly convergent, and the computation succeeds only because much analytical information is used. The motion turns out to be nearly damped harmonic except in an initial and a final stage.

W.25   Sequel to W.3, W.12 and W.14a. Computation of virtual mass and damping for oblique waves near a half-immersed circular cylinder. The expansion of W.23 is used; the coefficients are determined as in W.3. The results are checked against analytical asymptotics, and the computation would not have succeeded without these checks. (A complete set of asymptotic results has been obtained by plausible arguments. The sub-set used here has been rigorously established by my student M.W. Green.)

W.26   Davis has shown by means of a lengthy calculation that, for two-dimensional oscillations in a canal of width $2a$, the m-th eigenvalue has the form

$$N_m = \frac{1}{2}m - \frac{\lambda_1 + \lambda_2}{4m} + O\left(\frac{1}{m}\right),$$

where λ_1/a and λ_2/a are the curvatures of the bounding cross-sectional curve C at its vertical intersections with the free surface. Here the same result is obtained more simply.

W.27   Sequel to W.22. It is known that head seas cannot travel without deformation along a horizontal cylinder of full constant cross-section. Calculations are given which indicate that the waves are refracted away from the axis of the cylinder. Similar refraction effects are found for waves generated by a pulsating source on the cylinder, and also for the Kelvin wave pattern generated by a long ship of nearly constant cross-section moving with constant speed in the axial direction.

W.28   This is a sequel to W.19. A half-immersed circular cylinder is making vertical oscillations on water of finite constant depth. The virtual-mass and damping coefficients are studied in the limit as the wave-length tends to infinity. It is found that the virtual-mass coefficient tends to infinity, and the amplitude ratio ultimately varies as (frequency)^2. The computational difficulties which arise at long wave-lengths are analysed.

W.29   A sequel to W.22 and W.27. According to W.27, head seas are refracted away from the axis of a horizontal cylinder of full constant cross-section. According to W.22, head seas are refracted towards a ship of thin wedge-like cross-section. It was assumed in W.27 that the wave number Ka is neither small nor large. In W.29 the calculations of W.27 are repeated for small Ka. The asymptotic expansions must then be replaced by uniformly asymptotic expansions. It is shown that refraction becomes significant at distances $> x_0$ where $Ka(Kx_0)^{\frac{1}{2}} = 1$. The thin-ship calculation, on the other hand, applies only at distances $\ll x_0$. Thus the calculation of W.22 and W.27 are not really inconsistent, and the refraction is always away from the ship.

W.30   A sequel to B.4 and B.5. It is shown how irregular frequencies can be eliminated by placing additional wave singularities inside the cylinder. The argument is similar to the argument for sound waves given in B.4 and B.5 but involves the expansion of a wave-source in terms of multipole potentials (here given for the first time).

W.31   A horizontal cylindrical body moves with constant velocity in the horizontal direction normal to its axis, near the free surface of a frictionless fluid. If the free-surface condition is linearized we obtain the Kelvin-Neumann problem. (The linearization is not physically permissible near the body, nevertheless this problem has been much studied.) Here the half-immersed circular cylinder is studied. There are arguments which suggest that in the corners the velocity potential must be strongly singular, but it is shown here that a unique velocity potential exists which has bounded velocities in the corners and at infinity (except possibly at a set of irregular frequencies). Similar results hold for other cross-sections. It is suggested that the potential is logarithmically infinite at infinity, but Katsuo Suzuki shows in the discussion that it is in fact bounded, and this is confirmed by an independent calculation. The physical significance of this least singular potential is still unclear.

W.32   This is a sequel to W.16. In the linearized theory of wave resistance a ship in steady forward motion is represented as a distribution of Kelvin sources over the mid-plane of the ship. The velocity potential of the Kelvin source can be written in a form involving a double integral but in 1964 Bessho gave a simpler alternative form involving only single integrals. Although this form would be expected to have many advantages in analytical work it is rarely quoted in the literature. Bessho's mathematical argument was given in outline rather than in detail, and at least one important step is difficult to justify. In the present paper Bessho's work is re-examined, and it is shown that his results are in fact correct.

W.33   There is as yet no satisfactory uniqueness theory in the linearized theory of time-periodic water waves. Thus it is not known whether trapping modes (bound states of finite energy) can exist in domains which are unbounded at infinity. (They can exist in canals of finite width, see W.7.) Partial results are known; for example, W.6 gives a uniqueness proof for the submerged circular cylinder in infinite depth. In 1950 it was proved by F. John that uniqueness holds for bodies which intersect the free surface orthogonally in a curve C and lie inside the vertical cylinder through C. In the present work the method of F. John is generalized (in two dimensions only) to cover a wider class of immersed and submerged bodies.

W.34   This is the Eighth Georg Weinblum Memorial Lecture, delivered in Hamburg in November 1985 and in Washington in April 1986. It is an expository lecture, concerned with the calculation of waves generated by oscillating bodies without forward speed. For bodies of arbitrary shape the method of integral equations is available but for certain simple shapes the method of multipoles is much more effective. (This was first used for the half-immersed circular cylinder, in W.3). In fact there are many ways of representing velocity potentials, and it is explained why (depending on circumstances) some of them are more advantageous than others.

W.35   A sequel to W.7 in which it was shown that a trapping mode exists for a sufficiently small submerged circular cylinder. D.S. Jones (1953) used the theory of unbounded operators in unbounded domains to show the existence of a trapping mode for arbitrary submerged cylinders with a vertical plane of symmetry. Here a much simpler treatment is given which does not require symmetry, and which uses only the theory of bounded symmetric linear operators together with Kelvin's minimum-energy theorem of classical hydrodynamics.

W.36   The Kelvin ship-wave source is important in the mathematical theory of the wave resistance of ships but its velocity potential is difficult to evaluate numerically. In particular, the integral

$$F(x, \rho, \alpha) = \int_{-\infty}^{\infty} \exp\left\{-\frac{1}{2}\rho \cosh\left(2u - i\alpha\right)\right\} \cos\left(x \cosh u\right) du$$

in the source potential is difficult to evaluate when x and ρ are positive and small, and when $-\frac{1}{2}\pi \leq \alpha \leq \frac{1}{2}\pi$. In this work we are concerned with the asymptotic expansion of this integral when $x^2/4\rho$ is large while x and ϕ are not large, for which case the asymptotic expansion

$$F(x, \rho, \alpha) \sim -\pi I_0(\frac{1}{2}\rho)Y_0(x) - 2\pi \sum_{m=1}^{\infty} I_m(\frac{1}{2}\rho) Y_{2m}(x) \cos m\alpha$$

in terms of Bessel functions was proposed by Bessho (1964). This expansion has recently been shown to have great computational advantages but has never been proved. (The standard asymptotic theory of integrals, based on Watson's lemma, is not applicable, and the expansion is not of standard form.) In this paper it is shown that the expansion is valid except near $\alpha = \pm\frac{1}{2}\pi$ where an additional term is needed.

W.37   The same integral as in W.36 is treated. A convergent expansion is obtained, valid for small x and ρ. It is shown that

$$\begin{aligned} F(x, \rho, \alpha) &= \frac{1}{2}f(x, \rho, \alpha) + \frac{1}{2}f(x, \rho, -\alpha) \\ &+ \frac{1}{2}f(-x, \rho, \alpha) + \frac{1}{2}f(-x, \rho, -\alpha), \end{aligned}$$

where

$$\begin{aligned} f(x, \rho, \alpha) &= P_0(x, \rho e^{-i\alpha}) \sum g_m(x, \rho e^{i\alpha})c_m(x, \rho e^{i\alpha}) \\ &+ P_1(x, \rho e^{-i\alpha}) \sum g_m(x, \rho e^{-i\alpha})b_m(x, \rho e^{-i\alpha}) \\ &+ \sum g_m(x, \rho e^{i\alpha})a_m(x, \rho e^{-i\alpha}). \end{aligned}$$

In this expression each of the functions g_m, a_m, b_m and c_m satisfies a simple three-term recurrence relation and tends rapidly to 0 for small x and ρ when $m \to \infty$, and the functions P_0 and P_1 are simply related to the parabolic cylinder functions $D_\nu(\xi)$ and their derivatives $(\partial/\partial\nu)D_\nu(\xi)$ (for $\nu = 0$ and $\nu = 1$ respectively), where $\xi = -ix(2\rho)^{-\frac{1}{2}}e^{\frac{1}{2}i\alpha}$.

W.38   An account of three problems: the propagation of ocean waves (O.2); the heaving circular cylinder (W.3 and W.14a); and the Kelvin ship wave pattern (W.16 and W.37).

12.9.2   Acoustics

B.1   Sequel to W.12, which was the first known case of a rigorous short-wave limiting argument.   In B.1 the problem is the two-dimensional exterior Neumann problem of short-wave acoustics. The integral equation is obtained by taking as a fundamental solution the potential of a source on the local circle of curvature. This is without doubt the most difficult piece of mathematics which I have so far undertaken.

The method used here has received a number of further developments, particularly in the work of Grimshaw, also in the United States and in Russia.

B.2   Sequel to B.1. Various extensions of the work in B.1. The circle of curvature is inconvenient and should be replaced by an ellipse of close contact. Consideration is given to the extension to three dimensions which seems to be very difficult and has not yet been successful. An alternative very ingenious method has since been developed by C. Morawetz. This uses an inequality with (apparently) no physical meaning, and when combined with ray theory it gives results in both two and three dimensions. These latter methods have not yet succeeded with water wave problems like W.12 and W.18.

B.3   The papers B.1, B.2 make use of the Watson transformation for the short-wave limit of the fundamental solution. Watson in effect transformed a slowly convergent series into a contour integral, and thence into a different series of which the first few terms exhibit the law of shadow formation and later terms are neglected. This neglect was justified for a circle in B.1, but for an ellipse some of the neglected terms are exponentially large as was shown by my student Leppington. The present work shows that the sum of these large terms is nevertheless negligible; Watson's transformed series should be regarded as a finite sum with a small remainder. The argument has since been used to clear up analogous difficulties in refraction theory, discovered by D.S. Jones. See also M.4.

B.4   The method of integral equations is the most familiar method of proving existence theorems for the Helmholtz equation of acoustics. The wave potentials are expressed as surface distributions of wave sources (for the Neumann problem) or wave dipoles (for the Dirichlet problem). By a wave source is meant the free-space wave source. The source and dipole strengths for the exterior potentials are found to be solutions of Fredholm integral equations of the second kind which are, however, singular at a certain discrete set of frequencies corresponding to eigensolutions of the interior problems. The existence of exterior solutions at the expected frequencies can still be shown, but the proof involves a detailed and complicated study of the interior solutions. It is physically evident that this difficulty arises from the method of solution and not from the nature of the problem.

The present work shows this difficulty can be avoided. It is based on the well-known observation that the wave source of the theory need not be the free-space wave source but may be any fundamental solution defined in the exterior region. This idea is here illustrated in two dimensions for the case of a simply connected interior region. A certain fundamental solution is found explicitly which satisfies a dissipative boundary condition on a circle lying inside the interior region. For exterior problems the resulting theory is found to be formally identical with the usual theory, except that the integral equations are now non-singular at all frequencies. A similar construction is applicable in three dimensions.

B.5   This is a sequel to B.4. The fundamental solution described in B.4 satisfies a dissipative boundary condition on an interior circle. It has the advantage of removing all real irregular frequencies but the disadvantage of involving an infinite series of interior multipoles. It was shown by D.S. Jones that a finite number of irregular frequencies can be removed by adding a finite series of interior multipoles to the fundamental solution. Here a different and much shorter proof of one of Jones's principal results is given.

12.9.3 Asymptotics of integrals

M.1   (Motivated by the problem afterwards published in W.16). Consider an integral

$$\int g(z, \alpha) \exp\{Nf(z, \alpha)\} \, dz$$

when $N \to \infty$. The ordinary method of steepest descents succeeds when the saddle points are well separated, and it fails if two or more saddle points tend to coincidence as the parameter approaches a critical value. The asymptotic expansion then involves Airy functions. This has been known for over 100 years, but all proofs were unsatisfactory, and all forms of the result were non-uniform. A completely satisfactory theorem is obtained here by means of a new variable u satisfying a cubic relation

$$f(z, \alpha) = \frac{1}{3}u^3 - u\xi(\alpha) + A(\alpha).$$

For a certain choice of ξ and A this $(3,3)$ correspondence breaks up into a $(1,1)$ branch and a $(2,2)$ branch. This crucial result was proved independently by Chester and Friedman, and by myself. My proof is the one using the Riemann surface. The theorem is valid in a small region of the α-plane near the critical value, a region which is independent of the large parameter N. (Extension to several nearly coincident saddle points and parameters is given in M.5).

M.2 It is shown that the region of validity of the theorem obtained in M.1 can be extended into wider regions of the α-plane which are usually unbounded.

M.3 Sequel to B.2. The rigorous treatment of short-wave asymptotics leads to the following problem. Given an integral equation

$$\phi(\theta; N) + \int K(\theta, \alpha; N)\phi(\alpha; N)\, d\alpha = F(\theta; N),$$

for what classes of kernels K can we assert that

$$\max_{\theta} |\phi(\theta; N)| < A \max_{\theta} |F(\theta; N)|,$$

at least for all sufficiently large N? In B.1 the kernel tended to 0 in some norm and thus has this property; but it is plausible that rapidly oscillating kernels also have this property. Here the kernels $c(\theta, \alpha) \exp(iN \mid \theta - \alpha \mid)$ and $c(\theta, \alpha) \cos N(\theta - \alpha)$ are studied, where $c(\theta, \alpha)$ is sufficiently smooth. The first of these has the property, the second does not. This looks hopeful but no application to short-wave asymptotics has yet been made.

M.4 Sequel to B.3. The ordinary method of steepest descents involves a global problem: Given an initial element of a path of steepest descent, in which valley at ∞ does this path end? A general solution is not to be expected. It is here shown that the problem can be reduced to a problem of conformal mapping, and that a solution can in suitable cases be obtained in this way. (B.3 is a rather complicated example).

M.5 Sequel to M.1. The integral is of the form

$$\int g(z, \alpha_1, \ldots, \alpha_\ell) \exp\{N f(z, \alpha_1, \ldots, \alpha_\ell)\}\, dz,$$

with ℓ complex parameters and m nearly coincident saddle points. As in M.1, it is sufficient to consider the special case when f is a polynomial of degree $m + 1$. The asymptotic expansion involves generalized Airy functions and their derivatives; the coefficient functions are determined by Bleistein's method. The remainder after n terms is small provided that a certain inequality is satisfied for sufficiently large N, and the proof of this inequality is the most difficult part of the problem. The proof given here is valid for any values of ℓ and m. The region of validity of the expansion does not shrink to zero when $N \to \infty$.

M.6 This is a sequel to M.5. The double integral is

$$\iint G(x,y;\alpha,\beta,\gamma)\,e^{iN\Phi(x,y)}\,dx\,dy,$$

where

$$\Phi(x,y) = \frac{x^3}{3} + \frac{y^3}{3} + \alpha x + \beta y + \gamma xy,$$

involving two complex variables x and y, three complex parameters α, β, γ, and a large real parameter N. This integral occurs in optics and in theoretical chemistry. When α, β and γ tend to 0 there are four nearly coincident saddle points. The method of M.5 is generalized to give a complete asymptotic expansion. It is found that the principal difficulty lies in determining suitable surfaces of integration; in particular, surfaces of steepest descent cannot be defined in an invariant manner. Nevertheless results can be obtained for the most interesting case when α, β and γ are small.

M.7   The Hilbert transform of the function $f(t)$ is the integral

$$Hf(x) = \int_0^\infty \frac{f(t)}{t-x}\,dt = \lim_{\epsilon \to 0}\left(\int_0^{x-\epsilon} + \int_{x+\epsilon}^\infty\right)\frac{f(t)}{t-x}\,dt;$$

we are concerned with its behaviour when $x \to \infty$. The usual method of Mellin transforms is not applicable when $f(t)$ is oscillatory for large t,

$$f(t) \sim \sum_1^\infty \frac{a_n}{t^n} + \cos\omega t \sum_1^\infty \frac{A_n}{t^n} + \sin\omega t \sum_1^\infty \frac{B_n}{t^n},$$

but an alternative method (due to Wong) can be used to show that

$$Hf(x) \quad \sim \quad \sum_1^\infty \frac{d_n}{x^n} - \log x \sum_1^\infty \frac{a_n}{x^n}$$
$$- \quad \pi\sin\omega x \sum_1^\infty \frac{A_n}{x^n} + \pi\cos\omega x \sum_1^\infty \frac{B_n}{x^n},$$

where d_n involves an integral with a discontinuous integrand. Here it is shown that in fact

$$d_k = \lim_{\rho \to k}\left(M(\rho) + \frac{a_k}{\rho - k}\right)$$

where $M(\rho)$ is the Mellin transform of $f(t)$. The case $f(t) = J_0^2(t)$ is treated as an example.

M.8   This is a sequel to M.5. Asymptotic expansions for the Legendre functions $Q_n^{-m}(\cosh z)$ and $P_n^{-m}(\cosh z)$ for fixed m and large n have been derived by Olver, who used the differential equation. When z is small, these involve Bessel functions of order m. Here the same expansions are derived from integral expressions for the Legendre functions. The method is based on the observation that the asymptotic expansion for Q_n^{-m} involves the same coefficient functions as the asymptotic expansion for P_n^{-m}. It is shown that efficient uniform error bounds for these can be deduced from simpler non-uniform bounds by applying the maximum-modulus theorem.

M.9 The integral is

$$\varepsilon \int_0^\infty e^{-\varepsilon^2 z^2} \cos\{N\varepsilon^{\frac{1}{2}}(z \tanh z)^{\frac{1}{2}}\}\, dz$$

(with a large real parameter N and a small real parameter ε), and it arises in certain initial-value problems in the theory of water waves. Its behaviour is non-uniform when $N \to \infty$ and $\varepsilon \to 0$. It is shown that several distinct uniformly asymptotic expansions can be obtained which each involve an infinite set of functions of the combination $N\varepsilon^{\frac{1}{2}}$. Certain related integrals are also treated.

M.10 A strong form of Watson's Lemma. The function $f(t)$ in the integral

$$I(N) = \int_0^\infty e^{-Nt} f(t)\, dt$$

(where N is a large positive parameter) is assumed to have a convergent expansion

$$f(t) = \sum_{m=0}^\infty f_m t^m \quad \text{when } 0 \le t \le R;$$

it is also assumed that

$$|f(t)| < A e^{Bt} \quad \text{when } R \le t < \infty,$$

where A and B are constants. As is well known, Watson's Lemma states that with these assumptions

$$I(N) \sim \sum_{m=0}^M f_m \frac{m!}{N^{m+1}} + O\left(\frac{1}{N^{m+2}}\right),$$

with an algebraically small remainder. It is here shown that

$$I(N) \sim \sum_{m \le Nr} f_m \frac{m!}{N^{m+1}} + O(e^{-Nr}) \quad \text{where } 0 \le r < R.$$

In this expression the remainder is exponentially small. The conventional form of Watson's Lemma can be deduced from this result.

M.11 An account, with less mathematical detail, of the work described in M.5 and M.8, with some remarks on the extension of the domain of validity.

12.9.4   Aerodynamics

A.1   Incompressible lifting-surface theory for swept-back wings.

A.2   The general theorems are mainly flow reversal theorems in subsonic and supersonic flow. This paper also contains a general uniqueness theorem for lifting-surface theory incorporating the Joukowsky condition. There are errors in some of the results involving moments.

12.9.5   Oceanography

O.2   Also O.1, O.4 and O.6. Consider a disturbance in mid-ocean localized in time and space. According to linearized theory (due to Cauchy and Poisson), in a wind-free area the disturbance spreads along group-velocity rays. An observer on shore sees long waves first, then shorter and shorter waves. From the variation of wave period with time of arrival he can infer the time and distance of the original disturbance. This is the model that is applied to the North Atlantic. (First proposed by me in an internal ARL report.) Incoming waves are frequency-analysed and traced back to their region of origin on a time-distance propagation-diagram. (Theoretical deep-water waves are used for the group velocity.) It is found that the propagation lines are closely correlated with weather information. Correlations with wind strength and direction are obtained, and in one incident it was found that waves had travelled from the Southern Ocean.

In my view the success of this experiment is principally due to N.F. Barber's ingenious method of frequency analysis. In his method about 120 harmonics could be found in about 20 minutes, a remarkable achievement at that time (1945). Barber's idea is described in O.6.

This work has been the basis of modern methods of ocean wave forecasting, a topic that cannot be reviewed here.

Papers O.1 and O.6 describe aspects of the frequency analysis. Paper O.4 is a reply to a criticism. Sequel: O.7.

O.3   The mass transport associated with Stokes waves cannot persist in a rotating system unless lateral forces are acting. This is shown very simply by use of the Bjerknes circulation theorem. This result has been rediscovered by use of perturbation theory.

O.4   (with M.S. Longuet-Higgins) A simple proof of Miche's result on bottom pressures under a standing wave. Frequency doubling takes place.

O.7   A lengthy review of theories of wave generation by wind, starting with Kelvin-Helmholtz. It became clear that none of the existing theories was adequate, and publication was quickly followed by the theories of Phillips and Miles, which have so greatly contributed to our knowledge though they may not be the final answer. This review I found more troublesome to write than any of my other work.

12.9.6 Papers by research students

The following published papers are based on Ph.D. dissertations written under the supervision of F. Ursell, or on work with research assistants.

Bartholomeusz, E.F. 1958 The reflexion of long waves at a step. *Proc. Camb. Phil. Soc.* **54**, 106–118.

Thorne, R.C. 1953 Multipole expansions in the theory of surface waves. *Proc. Camb. Phil. Soc.* **49**, 707–716.

Thorne, R.C. 1957 The asymptotic solution of differential equations with a turning point and singularities. *Proc. Camb. Phil. Soc.* **53**, 382–398.

Thorne, R.C. 1957 The asymptotic solution of linear second order differential equations in a domain containing a turning point and a regular singularity. *Phil. Trans. Roy. Soc.* **A249**, 585–596.

Thorne, R.C. 1957 The asymptotic expansions of Legendre functions of large degree and order. *Phil. Trans. Roy. Soc.* **A249**, 597–620.

Newman, J.N. 1961 A linearized theory for the motion of a thin ship in regular waves. *J. Ship Res.* **5**, 34–55.

Hutson, V.C.L. 1963 The circular plate condenser at small separations. *Proc. Camb. Phil. Soc.* **59**, 211–224.

Tuck, E.O. 1964 Some methods for flows past blunt slender bodies. *J. Fluid Mech.* **18**, 619–635.

Tuck, E.O. 1964 A systematic asymptotic expansion procedure for slender ships. *J. Ship Res.* **8**, 15–23.

Tuck, E.O. 1964 On line distributions of Kelvin sources. *J. Ship Res.* **8**, 45–52.

Holford, R.L. 1964 Short surface waves in the presence of a finite dock. I. *Proc. Camb. Phil. Soc.* **60**, 957–983.

Holford, R.L. 1964 Short surface waves in the presence of a finite dock. II. *Proc. Camb. Phil. Soc.* **60**, 985–1011.

Davis, A.M.J. 1965 Two-dimensional oscillations in a canal of arbitrary cross-section. *Proc. Camb. Phil. Soc.* **61**, 827–846.

Davis, A.M.J. 1969 Short surface waves in a canal: dependence of frequency on curvature. *Proc. Roy. Soc.* **A313**, 249–260.

Evans, D.V. 1968 The influence of surface tension on the reflection of water waves by a plane vertical barrier. *Proc. Camb. Phil. Soc.* **64**, 795–810.

Evans, D.V. 1968 The effect of surface tension on the waves produced by a heaving circular cylinder. *Proc. Camb. Phil. Soc.* **64**, 833–847.

Leppington, F.G. 1967 Radiation of short waves by a convex cylinder. *Quart. J. Mech. Appl. Math.* **20**, 107–125.

Leppington, F.G. 1967 Creeping waves in the shadow of an elliptic cylinder. *J. Inst. Math. Applns.* **3**, 388–402.

Leppington, F.G. 1968 On the short-wave asymptotic solution of a problem in acoustic radiation. *Proc. Camb. Phil. Soc.* **64**, 1131–1150.

Gregory, R.D. 1966 The attenuation of a Rayleigh wave in a half-space by a surface impedance. *Proc. Camb. Phil. Soc.* **62**, 811–827.

Gregory, R.D. 1967 An expansion theorem applicable to problems of wave propagation in an elastic half-space containing a cavity. *Proc. Camb. Phil. Soc.* **63**, 1341–1367.

Rhodes-Robinson, P.F. 1970 On the short-wave asymptotic motion due to a cylinder heaving on water of finite depth. Part I. *Proc. Camb. Phil. Soc.* **67**, 423–442.

Rhodes-Robinson, P.F. 1970 On the short-wave asymptotic motion due to a cylinder heaving on water of finite depth. Part II. *Proc. Camb. Phil. Soc.* **67**, 443–468.

Green, M.W. 1971 A problem connected with the oblique incidence of surface waves on an immersed cylinder. *J. Inst. Math. Applns.* **8**, 82–98.

Martin, J. 1974 Integrals with a large parameter and several nearly coincident saddle points; the continuation of uniformly asymptotic expansions. *Proc. Camb. Phil. Soc.* **76**, 211–231.

Sayer, P. 1980 The long-wave behaviour of the virtual mass in water of finite depth. *Proc. Roy. Soc.* *A***372**, 65–91.

Sayer, P. 1980 An integral-equation method for determining the fluid motion due to a cylinder heaving on water of finite depth. *Proc. Roy. Soc.* *A***372**, 93–110.

Martin, P.A. 1980 On the null-field equations for the exterior problems of acoustics. *Quart. J. Mech. Appl. Math.* **33**, 385–396.

Hulme, A. 1981 The potential of a horizontal ring of wave sources in a fluid with a free surface. *Proc. Roy. Soc.* *A***375**, 295–305.

Hulme, A. 1983 A ring-source/integral-equation method for the calculation of hydrodynamic forces exerted on floating bodies of revolution. *J. Fluid Mech.* **128**, 387–412.

Burrows, A. 1985 Waves incident on a circular harbour. *Proc. Roy. Soc.* *A***401**, 349–371.